# Synthesis Lectures on Engineering, Science, and Technology

The focus of this series is general topics, and applications about, and for, engineers and scientists on a wide array of applications, methods and advances. Most titles cover subjects such as professional development, education, and study skills, as well as basic introductory undergraduate material and other topics appropriate for a broader and less technical audience.

Rajan Chattamvelli · Ramalingam Shanmugam

# Generating Functions in Engineering and the Applied Sciences

Second Edition

Rajan Chattamvelli
School of Advanced Sciences
Vellore Institute of Technology
Tamil Nadu, India

Ramalingam Shanmugam
School of Health Administration
Texas State University
San Marcos, TX, USA

ISSN 2690-0300                    ISSN 2690-0327   (electronic)
Synthesis Lectures on Engineering, Science, and Technology
ISBN 978-3-031-21145-4           ISBN 978-3-031-21143-0   (eBook)
https://doi.org/10.1007/978-3-031-21143-0

This Springer imprint is published by the registered company Springer Nature Switzerland AG
The registered company address is: Gewerbestrasse 11, 6330 Cham, Switzerland

# Preface

This is an introductory book on generating functions (GFs) and their applications. It discusses commonly encountered GFs in engineering and applied sciences, such as ordinary generating functions (OGF), exponential generating functions (EGF), and probability generating functions (PGF). Some new GFs like Pochhammer GFs for both rising and falling factorials, Dirichlet GF, Lambert GF, Hurwitz GF, Mittag-Lefler GF, etc. are introduced in Chap. 1. Common operations on GFs like addition, scalar multiplication, shifting, linear combinations, convolutions, differentiation and integration, etc. are discussed in Chap. 2. Two novel GFs called "mean deviation generating function" (MDGF), and "survival function generating function" (SFGF) are introduced in Chap. 3. Chapter 4 discusses a large number of applications in several areas including analysis of algorithms, bioinformatics, combinatorics, genomics, geometry, graph theory, management, number theory, polymer chemistry, reliability, statistics, structural engineering, etc.

Some background knowledge on GFs is often assumed for courses in analysis of algorithms, advanced data structures, graph theory, etc. These are usually provided by either a course on "discrete mathematics" or "introduction to combinatorics". However, GFs are also used in automata theory, bioinformatics, partial differential equations (PDE), digital signal processing (DSP), number theory, polymer chemistry, structural engineering, and so on. Students of these courses may not have exposure to discrete mathematics or combinatorics. This book is written in such a way that even those who do not have prior knowledge can easily follow through the chapters, and apply the lessons learned in their respective disciplines. The purpose is to give a broad exposure to commonly used techniques of combinatorial mathematics, highlighting applications in a variety of disciplines.

Any suggestions for improvement will be highly appreciated. Please send your comments to rajan.chattamvelli@vit.ac.in, and will be incorporated promptly.

Tamil Nadu, India
San Marcos, USA

Rajan Chattamvelli
Ramalingam Shanmugam

# Contents

# About the Authors

**Rajan Chattamvelli** is a professor in the school of advanced sciences at Vellore Institute of Technology, Tamil Nadu. He has published more than 20 research articles in international journals of repute, and at various conferences. His research interests are in computational statistics, design of algorithms, parallel computing, cryptography, data mining, machine learning, combinatorics, and bigdata analytics. His prior assignments include Denver Public Health, Colorado; Metromail Corporation, Lincoln, Nebraska; Frederick University, Cyprus; Indian Institute of Management; Periyar Maniammai University, Thanjavur; and Presidency University, Bangalore.

**Ramalingam Shanmugam** is a honorary professor in the school of Health Administration at Texas State University. He is the editor of the journals: *Advances in Life Sciences*, *Global Journal of Research and Review*, *International journal of research in Medical Sciences*, and book-review editor of the *Journal of Statistical Computation and Simulation*. He has published more than 200 research articles and 120 conference papers. His areas of research include theoretical and computational statistics, number theory, operations research, biostatistics, decision-making, and epidemiology. His prior assignments include University of South Alabama, University of Colorado at Denver, Argonne National Labs, Indian Statistical Institute, and Mississippi State University. He is a fellow of the International Statistical Institute.

# Abbreviations

| | |
|---|---|
| CDF | Cumulative Distribution Function |
| CDFGF | Cumulative Distribution Function GF |
| CGF | Cumulant Generating Function |
| CGP | Chain-Growth Polymerization |
| ChF | Characteristic Function |
| CHN | Central Harmonic Number |
| CMGF | Central Moment Generating Function |
| DGF | Dirichlet Generating Function |
| EGF | Exponential Generating Function |
| FCGF | Factorial Cumulant Generating Function |
| FMGF | Factorial Moment Generating Function |
| FPGF | Falling Pochhammer Generating Function |
| GP | Geometric Progression |
| HGF | Hurwitz Generating Function |
| IGF | Information Generating Function |
| IID | Independently and Identically Distributed |
| LGF | Lambert Generating Function |
| LogGF | Logarithmic Generating Function |
| MDGF | Mean Deviation Generating Function |
| MGF | Moment Generating Function |
| MLGF | Mittag-Lefler Generating Function |
| OGF | Ordinary Generating Function |
| PGF | Probability Generating Function |
| PMF | Probability Mass Function |
| RoC | Radius of Convergence |
| RPGF | Rising Pochhammer Generating Function |
| SF | Survival Function |
| SFGF | Survival Function Generating Function |

| SGP | Step-Growth Polymerization |
| SIR | Susceptible, Infected, Recovered |
| VGF | Variance Generating Function |
| WOGF | Weighted Ordinary Generating Function |

# Types of Generating Functions

**1**

A generating function (GF) is always derived from a well-ordered sequence as an algebraic equation. This chapter introduces the basic concepts on generating functions (GFs) like existence, uniqueness, radius of convergence, dummy variables, etc. It then describes the ordinary and exponential GFs, which are the most widely used in applied science fields. Some special GFs like Pochhammer GF, Dirichlet GF, Mittag-Lefler GF, Hurwitz GF, Lambert GF, logarithmic GF, auto-covariance GF, information GF, etc. are also discussed. The OGF of partial sums of a sequence is also derived, which is used to find the GF of harmonic numbers, and to get new GFs in Chap. 3.

## 1.1 Introduction

Generating function is a mathematical technique to concisely represent a known ordered sequence into a simple algebraic function. In essence, it takes a sequence as input, and produces a continuous function in one or more dummy (arbitrary) variables as output. A sequence is an ordered succession of elements which may be finite or infinite. A series is always associated with a *numeric* sequence and is the sum of the successive terms in the parent sequence. However, the input to a GF can be real or complex numbers, matrices, symbols, functions, or variables including random variables. The input depends on the area of application. For example, it is amino acid symbols in genomics or protein sequences in bio-informatics, random variables in statistics, digital signals in signal processing, number of infected persons in a population in epidemiology, etc. The output can contain unknown parameters, if any.

© The Author(s), under exclusive license to Springer Nature Switzerland AG 2023

R. Chattamvelli and R. Shanmugam, *Generating Functions in Engineering and the Applied Sciences*, Synthesis Lectures on Engineering, Science, and Technology,
https://doi.org/10.1007/978-3-031-21143-0_1

### 1.1.1  Origin of Generating Functions

Generating functions were introduced by the French mathematician Abraham de Moivre circa 1730, whose work originated in enumeration. It became popular with the works of Leonhard Euler, who used it to find the number of divisors of a positive integer (Chap. 4). The literal meaning of *enumerate* is to count or reckon something. This includes persons or objects, activities, entities, or known processes. Counting the number of occurrences of a discrete outcome is fundamental in enumerative combinatorics. For example, it may be placing objects into urns, selecting entities from a finite population, arranging objects linearly or circularly, etc. Thus, GFs are very useful to explore discrete problems involving finite or infinite sequences.

**Definition 1.1** (*Definition of Generating Function*) A GF is a simple and concise expression in one or more dummy variables that captures the coefficients of a finite or infinite sequence, and generates a quantity of interest using calculus or algebraic operations, or simple substitutions.

GFs are used in engineering and applied sciences to generate different quantities with minimal work. As they convert practical problems that involve sequences into the continuous domain, calculus techniques can be employed to get more insight into the problems or sequences. They are used in quantum mechanics, mathematical chemistry, population genetics, bio-informatics, vector differential equations, integral equations, epidemiology, stochastic processes, etc., and their applications in numerical computing include best uniform polynomial approximations, and finding characteristic polynomials. Average complexity of computer algorithms are obtained using GFs that have closed form (Chap. 4). This allows asymptotic estimates to be obtained easily. GFs are also used in polynomial multiplications in symbolic programming languages. In statistics, they are used in discrete distributions to generate probabilities, moments, cumulants, factorial moments, mean deviations, etc. (Chap. 3). The partition functions, Green functions, and Fourier series expansions are other examples from various applied science fields. They are known by different names in these fields. For example, $z$-transforms (also called $s$-transforms) used in digital signal processing (DSP), characteristic equations used in econometrics and numerical computing, canonical functions used in some engineering fields, and annihilator polynomials used in analysis of algorithms are all special types of GFs. Note, however, that the McClaurin series in calculus that expands an arbitrary continuous differentiable function around the point zero as $f(x) = f(0) + xf'(0) + (x^2/2!)f''(0) + \cdots$ is not considered as a GF in this text because it involves derivatives of a function.

## 1.1.2   Existence of Generating Functions

GFs are much easier to work with than a whole lot of numbers or symbols. This does not mean that any sequence can be converted into a GF. For instance, a random sequence may not have a nice and compact GF. Nevertheless, if the sequence follows a mathematical rule (like recurrence relation, geometric series, harmonic series), it is possible to express them compactly using a GF. Thus, it is extensively used in counting problems where the primary aim is to count the number of ways a task can be accomplished, if other parameters are known. A typical example is in counting objects of a finite collection or arrangement. Explicit enumeration is applicable only when the size (number of elements) is small. Either computing devices must be used, or alternate mathematical techniques like combinatorics, graph theory, dynamic programming, etc., employed when this size becomes large. In addition to the size, the enumeration problem may also involve one or more parameters (usually discrete). This is where GFs become very helpful. Many useful relationships among the elements of the sequence can then be derived from the corresponding GF. If the GF has closed form, it is possible to obtain a closed form for its coefficients as well. They are powerful tools to prove facts and relationships among the coefficients. Although they are called generating "functions," they are not like mathematical functions that have a domain and range. However, most of the GFs encountered in this book are polynomials in the dummy variable.

GFs allow us to describe a sequence in as simple a form as possible. Thus, a GF encodes a sequence in compact form, instead of giving the $n^{th}$ term as output. Depending on what we wish to generate, there are different GFs. For example, moment generating function (MGF) generates moments of a population, and probability generating function (PGF) generates corresponding probabilities. These are specific to each distribution. An advantage is that if the MGF of an arbitrary random variable $X$ is known, the MGF of any linear combination of the form $a * X + b$ can be derived easily. This reasoning holds for other GFs too.

### Convergence of Generating Functions

Convergence of the GF is assumed in some fields for limited values of the parameters or dummy variables. For example, GF associated with real-valued random variables assumes that the dummy variable $t$ (defined below) is in the range $[0, 1]$, whereas for complex-valued random variables the range is $[-1, +1]$. Similarly, the GF used in signal processing assumes that $t \in [-1, +1]$. As the input can be complex numbers in some signal processing applications, the GF can also be complex. A *formal power series* implies that nothing is assumed on the convergence. As shown below, the sequence $1, x, x^2, x^3, \ldots$ has ordinary generating function (OGF) $F(x) = 1/(1 - x)$, which converges in the region $0 < x < 1$, while a sub-sequence $1, 0, x^2, 0, x^4, \ldots$ has OGF $G(x) = 1/(1 - x^2)$, which converges for

$-1 < x < 1$. The region within which it converges is called the Radius of Convergence (RoC). Thus the RoC of a sequence is the set of bounded values within which it will converge. The RoC of sub-sequences could differ from the RoC of original sequence. Obviously, $(1, 1, \ldots, \infty)$ has OGF $1/(1 - x)$, so that the RoC is $0 < x < 1$. Similarly, $(c, c, c, \ldots)$ is generated by $1/(1 - cx)$ so that the RoC is $0 < x < 1/c$. An OGF being weighted by $x^k$ will converge if all terms beyond a value (say $R$) is upper-bounded. Such a value of $R$, which is a finite real number, is called its RoC. This is given by the ratio $1/R = Lt_{n \to \infty} |a_{n+1}/a_n|$, where vertical bars denote absolute value for real coefficients. It can easily be found by taking the ratio of two successive terms of the sequence, if the $n^{th}$ term is expressed as a simple mathematical function. The RoC of a sequence can give important information on the growth of the coefficients. For GFs defined in the complex plane as $F(z) = \sum_{n=-\infty}^{\infty} a_n z^n$, the region $R_1 < z < R_2$ where it converges is called the annulus. F(z) is analytic (holomorphic) if $|z| < $ R and has at least one singularity on the circle $|z| = $R where R is the RoC.

If $R_1 = \lim \sup_{k \to \infty} |a_{-k}|^{1/k}$ and $R_2 = 1/\lim \sup_{k \to \infty} |a_k|^{1/k}$ are such that $R_1 < R_2$, then $F(z)$ converges. In all other cases $F(z)$ is either nonexistent or diverges. Convergence of a GF is often ignored in this book because it is rarely evaluated if at all, and that too for dummy variable values 0 or 1 (an exception appears in Chap. 4).

## 1.2   Notations and Nomenclatures

A finite sequence will be denoted by placing square-brackets around them, assuming that each element is distinguishable. An English alphabet will be used to identify it when necessary. An infinite sequence will be denoted by simple brackets, and named using Greek alphabet. Thus, $S = [a_0, a_1, a_2, a_3, a_4]$ denotes a finite sequence, whereas $\sigma = (a_0, a_1, a_2, a_3, \ldots a_n, \ldots)$ denotes an infinite sequence. The summation sign ($\sum$) will denote the sum of the elements that follow. The index of summation will either be explicitly specified, or denoted implicitly. Thus, $F(t) = \sum_{k=0}^{\infty} a_k t^k$ is an explicit summation in which $k$ varies from 0–$\infty$ whereas $F(t) = \sum_{k \geq 0} a_k t^k$ implicitly assumes that the summation is varying from 0 to an upper limit (which will be clear from context). The notation $[t^n]F(t)$ for $n \in Z$ denotes $a_n$, the coefficient of $t^n$ in the OGF $F(t)$. Similarly, $[t^n/n!]H(t)$ is the coefficient $a_n$ in exponential GF (EGF) $H(t)$. Thus $[t^0]F(t) = a_0$, is the constant term, irrespective of whether $F(t)$ is an OGF or EGF. Note that this may differ for other GFs discussed below (like Dirichlet GF). An assumption that $n$ is a power of 2 may simplify the derivation in some cases (see Chap. 4). Random variables will be denoted by uppercase letters, and particular values of them by lowercase letters.

### 1.2.1 Rising and Falling Factorials

This section introduces a particular type of the product form presented above. These expressions are useful in finding factorial moments of discrete distributions, whose probability mass function (PMF) involves factorials or binomial coefficients. In the literature, these are known as Pochhammer's notation for rising and falling factorials. This will be explored in subsequent chapters.

1. Rising Factorial Notation
   Factorial products come in two flavors. In the rising factorial, a variable is incremented successively in each iteration. This is denoted as

   $$x^{(j)} = x * (x+1) * \cdots * (x+j-1) = \prod_{k=0}^{j-1}(x+k) = \frac{\Gamma(x+j)}{\Gamma(x)}. \qquad (1.1)$$

   Derivative of $x^{(j)}$ can be expressed in terms of $\psi(s) = (\partial/\partial x) \log(\Gamma(x))$ as $(\partial/\partial x) x^{(j)} = x^{(j)}[\psi(s+j) - \psi(s)]$. The value evaluated at $x = -j$ is a multiple of harmonic numbers as $(\partial/\partial x) x^{(j)}|_{x=-j} = -j!H_j$.

2. Falling Factorial Notation
   In the falling factorial, a variable is decremented successively at each iteration. This is denoted as

   $$x_{(j)} = x * (x-1) * \cdots (x-j+1) = \prod_{k=0}^{j-1}(x-k) = \frac{x!}{(x-j)!} = j!\binom{x}{j} = \frac{\Gamma(x+1)}{\Gamma(x-j+1)}, \qquad (1.2)$$

   where $\binom{x}{j}$ are the binomial coefficients (sometimes called central binomial coefficients). Some authors use the notation $< x >_j$ for falling factorial.

3. Combined Rising and Falling Factorial Notation
   A variable can be incremented and decremented successively at each iteration. This is denoted as

   $$(x-j)^{(2j+1)} = (x-j) * (x-j+1) * \cdots (x-1) * x * (x+1) * \cdots (x+j) = \prod_{k=-j}^{j}(x+k). \qquad (1.3)$$

Writing Eq. (1.1) in reverse gives us the relationship $x^{(j)} = (x+j-1)_{(j)}$. Similarly, writing Eq. (1.2) in reverse gives us the relationship $x_{(j)} = (x-j+1)^{(j)}$.

## 1.2.2   Dummy Variable

An OGF is denoted by $G(t)$, where $t$ is the dummy (arbitrary) variable. It can be any variable you like, and it depends on the field of application. Quite often, $t$ is used as dummy variable in statistics, $s$ in signal processing, $x$ in genetics, bio-informatics, and orthogonal polynomials, $z$ in analysis of algorithms, and $L$ in auto-correlation, and time-series analysis.[1] An EGF is denoted by $H(t)$. As GFs in some fields are associated with variables (like random variables in statistics), they may appear before the dummy variable. For example, $G(x, t)$ denotes an OGF of a random variable $X$. Some authors use $x$ as a subscript as $G_x(t)$. These dummy variables assume special values (like 0 or 1) to generate a quantity of interest. Thus, uniform and absolute convergence at these points are assumed. The GFs used in statistics can be finite or infinite, because they are defined on (sample spaces of) random variables, or on a sequence of random variables. Note that the GF is also defined for bivariate and multivariate data. Bivariate GF has two dummy variables. This book focuses only on univariate GF. Although the majority of GFs encountered in this book are of "sum-type" (additive GF), there also exist GFs of "product-type." One example is the GF for the number of partitions of an integer (Chap. 4). These are called multiplicative GF. The GF for the number of partitions of $n$ into distinct parts is $p(t) = \prod_{n=1}^{\infty}(1 + t^n)$.

## 1.3    Ordinary Generating Functions

Let $\sigma = (a_0, a_1, a_2, a_3, \ldots a_n, \ldots)$ be a sequence of bounded numbers. Then the power series

$$F(x) = \sum_{n=0}^{\infty} a_n x^n \qquad (1.4)$$

is called the OGF of $\sigma$. Here $x$ is a dummy variable, $n$ is the indexing variable (indexvar) and $a_n'$s are known constants. For different values of $a_n$, we get different OGF. The set of all possible values of the indexing variable of the summation in (1.4) is called the "index set." If the sequence is finite and of size $n$, we have a polynomial of degree $n$. Otherwise it is a power series. Such a power series has an inherently powerful enumerative property. As they lend themselves to algebraic manipulations, several useful and interesting results can be derived from them as shown in Chap. 2. There exist a one-to-one correspondence between a sequence, and its GF. This can be represented as

---

[1] Analysis of algorithms is used to either select an algorithm with minimal run time from a set of possible choices, or minimum storage or network communication requirements. As the attribute of interest is time, the recurrences are developed as $T(n)$ where $n$ is the problem size. This is one of the reasons for choosing a dummy variable other than $t$.

$$(a_0, a_1, a_2, a_3, \ldots, a_n, \ldots) \Longleftrightarrow \sum_{n=0}^{\infty} a_n x^n. \tag{1.5}$$

We get $F(x) = (1 - x)^{-1}$ when all $a_n = 1$ in (1.5); and $(1 + x)^{-1}$ when $a_n = -1$ for $n$ odd, and $a_n = +1$ for $n$ even. These are represented as

$$(1, 1, 1, 1, \ldots, 1, \ldots) \Longleftrightarrow (1 - x)^{-1}$$
$$(1, -1, 1, -1, \ldots, 1, -1, \ldots) \Longleftrightarrow (1 + x)^{-1}.$$

The OGF is $(x^n - 1)/(x - 1)$ when there are a finite number of 1's (say $n$ of them). Thus, $(1, 1, 1, 1, 1)$ has OGF $(x^5 - 1)/(x - 1)$. Similarly, we get $(1 - x^2)^{-1}$ when even coefficients $a_{2n} = +1$, and odd coefficients $a_{2n+1} = 0$ with resulting series $\{1, 0, 1, 0, \cdots\}$. The OGF of this is $1 + x^2 + x^4 + \cdots = \sum_{2|n} a_n x^n$ where $2|n$ is read as "2 divides n". This can also be written as $\{a_{c+np}\}, n = 0, 1, \cdots$, where $c$ is the initial index and $p$ the lag (both of which are integers). This corresponds to $F(x; c, p) = \sum_{n=0}^{\infty} a_{c+np} x^{c+np} = x^c \sum_{n=0}^{\infty} a_{c+np} (x^n)^p$. Note that $G(x; c, p) = \sum_{n=0}^{\infty} a_{c+np} x^n$ is a new GF in which the coefficients are from a subsequence of the original. Hence any subsequence can be represented in this notation. It is also helpful to represent left-shifted OGF as $G(x) = [F(x) - \sum_{n=0}^{m-1} a_n x^n]/x^m = \sum_{n \geq 0} a_{m+n} x^n$. Similarly, an OGF multiplied by a power of the dummy variable can be represented as $x^m * F(x) = \sum_{n \geq m} a_{n-m} x^n = a_0 x^m + a_1 x^{m+1} + \cdots$. For instance, $\{1, 0, 1, 0, \cdots\}$ is denoted simply as $\{a_{2n}\}$. The OGF of any finite number of non-overlapping subsequences can be found using the roots of unity (Chap. 4). The OGF is called the McClaurin series of the function on the left-hand side (LHS) when the right-hand side (RHS) has functional values as coefficients. In the particular case when the sum of all the coefficients is 1, it is called a PGF, which is discussed at length in Chap. 3.

**Problem 1.1** Prove that $\sum_{m|n} a_n x^n = 1/(1 - x^m)$ where $m|n$ is read as "m divides n", and $m < n$ are integers.

**Problem 1.2** Prove that $(F(x) - F(-x))/2 = \sum_{n=0}^{\infty} a_{2n+1} x^{2n+1} = \sum_{2|(n-1)} a_n x^n$.

Consider the set of all positive integers $P = (1, 2, 3, 4, 5, \ldots)$. This has OGF $1 + 2x + 3x^2 + 4x^3 + \cdots$. We know that $1/(1 - x) = 1 + x + x^2 + x^3 + x^4 + \cdots$. Differentiate both sides with respect to (w.r.t.) $x$ to get $-1/(1 - x)^2 * (-1) = 1 + 2x + 3x^2 + 4x^3 + \cdots$. Thus, the OGF is $1/(1 - x)^2$ or equivalently $(1 - x)^{-2}$. This can be represented as

**Table 1.1** Some standard generating functions

| Function | Series | Type | Notation |
|---|---|---|---|
| $1\pm x$ | $1\pm x$ | OGF | $(1,\pm 1,0,0,\ldots)$ |
| $1+2x+3x^2$ | $1+2x+3x^2$ | OGF | $(1,2,3,0,0,\ldots)$ |
| $\frac{1}{1-x}$ | $\sum_{k=0}^{\infty} x^k$ | OGF | $(1,1,1,1,\ldots)$ |
| $\zeta(x)$ | $\sum_{k=0}^{\infty} x^{-k}$ | DGF | $(1,1,1,1,\ldots)$ |
| $\frac{1}{1+x}$ | $\sum_{k=0}^{\infty}(-1)^k x^k$ | OGF | $(1,-1,1,-1,\ldots)$ |
| $\frac{1}{1-2x}$ | $\sum_{k=0}^{\infty} 2^k x^k$ | OGF | $(1,2,4,8,\ldots)$ |
| $\frac{1}{(1-x)^2}$ | $\sum_{k=0}^{\infty}(k+1)x^k$ | OGF | $(1,2,3,4,\ldots)$ |
| $\frac{1}{1-ax}$ | $\sum_{k=0}^{\infty} a^k x^k$ | OGF | $(1,a,a^2,a^3,\ldots)$ |
| $\frac{1}{1-x^2}$ | $\sum_{k=0}^{\infty} x^{2k}$ | OGF | $(1,0,1,0,\ldots)$ |
| $\exp(x)$ | $\sum_{k=0}^{\infty} x^k/k!$ | EGF | $(1,1,1,1,\ldots)$ |
| $\exp(ax)$ | $\sum_{k=0}^{\infty}(ax)^k/k!$ | EGF | $(1,a,a^2,a^3,\ldots)$ |

$$(1,2,3,4,\ldots,n,\ldots) \Longleftrightarrow (1-x)^{-2} = 1/(1-x)^2$$

$$(1,-2,3,-4,\ldots,n,-(n+1),\ldots) \Longleftrightarrow (1+x)^{-2} = 1/(1+x)^2.$$

See Table 1.1 for more examples, and Theorem 1.2 in Sect. 1.3.3.

### 1.3.1　Recurrence Relations

A recurrence relation is a rule that expresses a particular term of a sequence in terms of other elements. A forward recurrence is one that expresses the $n^{th}$ term in terms of preceding terms. These are algebraic equations with $n^{th}$ term on the LHS and an expression (like linear combination) on the RHS. One example is the Fibonacci recurrence $F_n = F_{n-1} + F_{n-2}$. There are in general two ways to represent the $n^{th}$ term of a sequence: (i) as a function of $n$, and (ii) as a recurrence relation between $n^{th}$ term and one or more of the previous terms. For example, $a_n = 1/n$ gives an explicit expression for $n^{th}$ term. Taking the ratio $a_n/a_{n-1} = (n-1)/n = (1-1/n)$ gives $a_n = (1-1/n)a_{n-1}$, which is a first-order linear recurrence relation (Chattamvelli and Jones 1995).

A recurrence relation can be defined either for a sequence, or on the parameters of a function (like a density function in statistics). It expresses the $n^{th}$ term in terms of one or more prior terms. First-order recurrence relations are those in which $n^{th}$ term is related to $(n-1)^{th}$ term. It is called a linear recurrence relation if this relation is linear. First-order linear recurrence relations are easy to solve, and it requires one initial condition. A second-order recurrence relation is one in which the $n^{th}$ term is related to $(n-1)^{th}$ and $(n-2)^{th}$

terms. It needs two initial conditions to ensure that the ensuing sequence is unique. One classical example is the Fibonacci numbers. In general, an $n^{th}$-order recurrence relation expresses the $n^{th}$ term in terms of $n-1$ previous terms. Consider the Bell numbers defined as $B_n = \sum_{k=0}^{n-1} \binom{n-1}{k} B_k$. This generates 1, 1, 2, 5, 15, 52, 203, etc. Recurrence relations are extensively discussed in Chap. 4).

## 1.3.2   Types of Sequences

There are many types of sequences encountered in engineering. Examples are arithmetic sequence, geometric sequence, exponential sequence, Harmonic sequence, etc. Alternating sequence is one in which the sign of the coefficients alter between plus and minus. This is of the form $(a_1, -a_2, a_3, -a4, \ldots)$. The sum of an alternating sequence may be convergent or divergent. Consider $\sum_{n=1}^{\infty} (-1)^{n+1} a_n$. Take $a_n = 1/n$ to get the alternating harmonic series $\sum_{n=1}^{\infty} (-1)^{n+1} (1/n) = 1 - 1/2 + 1/3 - 1/4 + \cdots$, which is convergent as the sum is $\log(2)$. Similarly, $\sum_{n=1}^{\infty} (-1)^{n+1}/(2n+1)$ converges to $\pi/4$.

A telescopic series is one in which all terms of its partial sum cancels out except the first or last (or both). Consider $\sum_{n=1}^{\infty} 1/[n(n+1)] = 1/2 + 1/6 + 1/12 + \cdots$. Use partial fractions to write $1/[n(n+1)] = 1/n - 1/(n+1)$ so that when the upper limit of the summation is finite, all terms cancel out except the first and last.

Consider the finite geometric sequence $[P, Pr, Pr^2, \ldots, Pr^n]$. It is called a geometric progression (GP) in mathematics and geometric sequence in engineering. It also appears in various physical, natural, and life sciences. As examples, the equations of state of gaseous mixtures (like van der Vaals equation, Virial equation), mean-energy modeling of oscillators used in crystallography and other fields, and geometric growth and deprecation models used in banking and finance can all be modeled as geometric series. Such series also appear in population genetics, insurance, exponential smoothing models in time-series, etc. Hartleb et al. (2020) used the GF method to investigate the free time-intervals between buyers at cash-register queues in supermarkets using various "gap distribution" functions.

In a GP, the first term is $P$, and rate of progression (called common ratio) is $r$. It is possible to represent the sum of any $k$ terms in terms of 3 variables, $P$ (initial term), $r$ (common ratio), and $k$. An interesting property of the geometric progression is that $a_{k-1} a_{k+1} = a_k^2$, i.e., the square of $k^{th}$ term is the product of the two terms around it. A geometric series can be finite or infinite. It is called an alternating geometric sequence if the signs differ alternatively. Let $S_n = [1, r, r^2, r^3, \ldots, r^n]$ denote a finite geometric sequence. Then the sum $\sum_{k=0}^{n} r^k = (r^{n+1} - 1)/(r - 1)$. Both the numerator and denominator are negative when $0 < r < 1$, so that we could write the RHS in the alternate form $(1 - r^{n+1})/(1 - r)$.

**1. OGF of alternating geometric sequence**

Find the OGF of alternating geometric sequence $(1, -a, a^2, -a^3, a^4, \ldots)$.

Consider the infinite series expansion of $(1 + ax)^{-1} = 1 - ax + a^2x^2 - a^3x^3 + \cdots$, which is the OGF of the given sequence. Here "$a$" is any nonzero constant.

**2. OGF of $a_n = 2^n + 3^n$**

Find the OGF of (i) $a_n = 2^n + 3^n$ and (ii) $a_n = 2^n + 1/3^n$.

The OGF is $G(x) = \sum_{n=0}^{\infty}(2^n + 3^n)x^n$. Split this into two terms and use the closed form formula for geometric series to get $G(x) = 1/(1 - 2x) + 1/(1 - 3x) = (2 - 5x)/[(1 - 2x)(1 - 3x)]$ which is the required GF. In the second case a similar procedure yields $G(x) = 1/(1 - 2x) + 1/(1 - x/3)$ which is the required OGF.

**3. OGF of odd positive integers**

Find the OGF of the sequence $1, 3, 5, 7, 9, 11, \ldots$

The $n^{th}$ term of the sequence is obviously $(2n + 1)$. Write $G(x) = \sum_{n=0}^{\infty}(2n + 1)x^n$. Split this into two terms, and take the constant 2 outside the summation to get $G(x) = 2\sum_{n=0}^{\infty}nx^n + \sum_{n=0}^{\infty}x^n$. The first sum is $2x/(1 - x)^2$ and second one is $1/(1 - x)$ so that $G(x) = 2x/(1 - x)^2 + 1/(1 - x)$. Take $(1 - x)^2$ as a common denominator and simplify to get $G(x) = (1 + x)/(1 - x)^2$ as the required OGF.

**4. OGF of perfect squares**

Find the OGF of all positive perfect squares $(1, 4, 9, 16, 25, \ldots)$.

Consider the expression $x/(1 - x)^2$. From Sects. 1.1.2–1.2 we see that this is the infinite series expansion

$$x/(1 - x)^2 = x + 2x^2 + 3x^3 + \cdots + nx^n + \cdots. \tag{1.6}$$

Differentiate both sides w.r.t. $x$ to get

$$(\partial/\partial x)\left(x/(1 - x)^2\right) = 1 + 2^2x + 3^2x^2 + \cdots + n^2x^{n-1} + \cdots. \tag{1.7}$$

If we multiply both sides by $x$ again, we see that the coefficient of $x^n$ on the RHS is $n^2$. Hence, $x\frac{\partial}{\partial x}(x/(1 - x)^2)$ is the GF that we are looking for. To find the derivative of $f = x/(1 - x)^2$, take log and differentiate to get[2]

---

[2] Note that the $n^{th}$ derivative of $1/(1 - x)$ is $n!/(1 - x)^{n+1}$ and that of $1/(1 - ax)^b$ is $(n + b - 1)!a^n/[(b - 1)!(1 - ax)^{b+n}]$.

$$(1/f)(\partial f / \partial x) = 1/x + 2/(1-x) = (1+x)/[x(1-x)] \tag{1.8}$$

from which $\frac{\partial f}{\partial x} = (1+x)/(1-x)^3$. Multiply this expression by $x$ to get the desired GF of perfect squares of all positive integers as $x(1+x)/(1-x)^3$. This can be represented symbolically as follows:

$$(1, 1, 1, 1, \ldots, n, \ldots) \Longleftrightarrow (1-x)^{-1} = 1/(1-x)$$

Derivative:  $(1, 2, 3, 4, \ldots, n, \ldots) \Longleftrightarrow (1-x)^{-2} = 1/(1-x)^2$

Rightshift:  $(0, 1, 2, 3, 4, \ldots, n, \ldots) \Longleftrightarrow x(1-x)^{-2} = x/(1-x)^2$

Derivative:  $(1, 4, 9, 16, \ldots, n^2, \ldots) \Longleftrightarrow \dfrac{\partial}{\partial x}(x/(1-x)^2) = (1+x)/(1-x)^3 \tag{1.9}$

Rightshift:  $(0, 1, 4, 9, 16, \ldots, n^2, \ldots) \Longleftrightarrow x(1+x)/(1-x)^3.$

Derivative:  $(1, 8, 27, 64, \ldots, n^3, \ldots) \Longleftrightarrow (1+4x+x^2)/(1-x)^4.$

Rightshift:  $(0, 1, 8, 27, 64, \ldots, n^3, \ldots) \Longleftrightarrow x(1+4x+x^2)/(1-x)^4.$

## 5. OGF of even sequence

Find OGF of the sequence $(2, 4, 10, 28, 82, \ldots)$.

Suppose we subtract 1 from each number in the sequence. Then we get $(1, 3, 9, 27, 81, 243, \ldots)$, which are powers of 3. Hence, above sequence is the sum of $(1, 1, 1, 1, \ldots) + (1, 3, 3^2, 3^3, 3^4, \ldots)$ with respective OGFs $1/(1-x)$ and $1/(1-3x)$. Add them to get $1/(1-x) + 1/(1-3x)$ as the answer.

## 1.3.3   OGF for Partial Sums

If the GF for any sequence is known in compact form (say $F(x)$), the GF for the sum of the first n terms of that particular sequence is given by $F(x)/(1-x)$. A consequence of this is that if we know the OGF of the sum of the first n terms of a sequence, we could get the original sequence by multiplying it by $(1-x)$. As multiplying by $(1-x)$ is the same as dividing by $1/(1-x)$ we could state this as follows:

**Theorem 1.1** (OGF of $G(x)/(1-x)$)  *Multiplying an OGF by $1/(1-x)$ results in the OGF of the partial sums of coefficients, and dividing by $1/(1-x)$ results in differenced sequence.*  □

**Proof** Let $G(x)$ be the OGF of the infinite sequence $(a_0, a_1, \ldots, a_n, \ldots)$. By definition $G(x) = a_0 + a_1 x + a_2 x^2 + \cdots + a_n x^n + \cdots$. Expand $(1-x)^{-1}$ as a power series $1 + x + x^2 + \cdots$, and multiply with $G(x)$ to get the RHS as $g(x) = (1-x)^{-1} * G(x) =$

$$(1 + x + x^2 + x^3 + \cdots)(a_0 + a_1 x + a_2 x^2 + \cdots + a_n x^n + \cdots) =$$

$$\left( \sum_{j=0}^{\infty} 1.x^j \right) \left( \sum_{k=0}^{\infty} a_k.x^k \right). \tag{1.10}$$

$\square$

Change the order of summation to get

$$\left( \sum_{k=0}^{\infty} \left( \sum_{j=0}^{k} a_j.1 \right) x^k \right) = \sum_{k=0}^{\infty} \left( \sum_{j=0}^{k} a_j \right) x^k = a_0 + (a_0 + a_1)x + (a_0 + a_1 + a_2)x^2 + \cdots.$$

$$\tag{1.11}$$

This is the OGF of the given sequence. This can be expressed as in (1.12) below.

$$(a_0, a_1, a_2, a_3, \ldots, a_n, \ldots) \iff F(x)$$

$$(a_0, a_0 + a_1, a_0 + a_1 + a_2, \ldots) \iff F(x)/(1-x). \tag{1.12}$$

This result has great usefulness. Computing the sum of several successive terms of a sequence appears in several fields of applied sciences. If the GF of such a sequence has closed form, it is a simple matter of multiplying the GF by $1/(1-x)$ to get the new GF for the partial sum; and multiplying by $1-x$ gives the OGF of differences. Assume that the OGF of a sequence $a_n$ is known in closed form. We seek the OGF of another sequence $d_n = a_n - a_{n-1}$. As this is a telescopic series, $\sum_{k=1}^{n} d_k = a_n - a_0$. It follows that $D(x) = \sum_n d_n x^n = (1-x)A(x)$, where $A(x)$ is the OGF of $a_n$'s.

Consider the problem of finding the sum of squares of the first $n$ natural numbers as $\sum_{k=1}^{n} k^2$. The OGF of $n^2$ was found in Page 10 as $x(1+x)/(1-x)^3$. If this is divided by $(1-x)$, we should have the OGF of $\sum_{k=1}^{n} k^2$. This gives the OGF as $x(1+x)/(1-x)^4$.

## 6. OGF of harmonic series $H_n$

Find the OGF of $H_n = 1 + 1/2 + 1/3 + \cdots + 1/n$.

The sequence $H_n = \sum_{k=1}^{n} 1/k$ is called the central harmonic number (CHN) or simply harmonic number. Summing in reverse gives $H_n = \sum_{k=0}^{n-1} 1/(n-k)$. Obviously, the partial sums $\{1, 1+\frac{1}{2}, 1+\frac{1}{2}+\frac{1}{3}, \cdots, \sum_{k=1}^{n} 1/k\}$ forms a divergent series asymptotically. The first few of them are $\{1, 3/2, 11/6, 25/12, \cdots\}$. It has several interesting applications in the analysis of algorithms. $H_1 = 1$ is the only integer in this sequence. All others are rational numbers. Although $H_n$ diverges, $H_n(\epsilon) = \sum_{k=1}^{n} 1/k^{1+\epsilon}$ converges for $\epsilon > 0$.

We know that $-\log(1-x) = \log(1/(1-x)) = x + x^2/2 + x^3/3 + x^4/4 + \cdots = \sum_{k=1}^{\infty} x^k/k$. We also know that when an OGF is multiplied by $1/(1-x)$, we get a new OGF in which the coefficients are the partial sum of the coefficients of the original one. Thus, the above sequence has OGF $G(x) = -\log(1-x)/(1-x)$. This result is used in the time complexity analysis of quicksort algorithm in Chap. 4. The EGF (Sect. 1.3.3) of CHN can be expressed in terms of incomplete gamma functions as $\sum_{n=1}^{\infty} H_n x^n/n! = \exp(x)[\ln(x) + \Gamma(0, x) + \gamma]$ where $\gamma = 0.577215665$ is Euler-Mascheroni constant. CHN is related to bino-mial coefficients as $H_n = \sum_{k=1}^{n} (-1)^{k+1} \binom{n}{k}/k$. For example, $H_4 = \binom{4}{1} - \binom{4}{2}/2 + \binom{4}{3}/3 - \binom{4}{4}/4 = 4$ -6/2+4/3-1/4 = 2.083333. An inverse identity exists as $\sum_{k=1}^{n} (-1)^{k+1} \binom{n}{k} H_k = -1/n$.

**Problem 1.3** If $0 < m < 1$, compute the sum $\sum_{n=0}^{\infty} H_n/m^n$.

**Problem 1.4** Find the OGF of the sequence $\{a_n = H_n/[n(n-1)]\}$.

**Problem 1.5** Find the OGF of $1 + 1/3 + 1/5 + \cdots + 1/(2n-1)$.

As another example, consider the sequence $(1/2, -\frac{1}{3}(1 + \frac{1}{2}), \frac{1}{4}(1 + \frac{1}{2} + \frac{1}{3}), \cdots)$. First form the convolution of $\log(1+x)$ and $1/(1+x)$ to get $\log(1+x)/(1+x) = x - x^2(1 + 1/2) + x^3(1 + 1/2 + 1/3) + \cdots$. Now integrate both sides to get the OGF as $(1/2)[\log(1 + x)]^2$. Similarly, $(1/2)[\log(1-x)]^2$ is the OGF of $(1/2, \frac{1}{3}(1 + \frac{1}{2}), \frac{1}{4}(1 + \frac{1}{2} + \frac{1}{3}), \cdots)$.

A direct application of the above result is in proving the combinatorial identity $\sum_{k=0}^{m} (-1)^k \binom{n}{k} = (-1)^m \binom{n-1}{m}$. First consider the sequence $(-1)^k \binom{n}{k}$, which has OGF $(1-x)^n$. According to the above theorem, the sum of the first $m$ terms has OGF $(1-x)^n/(1-x)$. Cancel out $(1-x)$ to get $(1-x)^{n-1}$, which is the OGF of the RHS. This proves the result.

**Theorem 1.2** If $F(t) = \sum_{k=0}^{\infty} a_k t^k$, then $G(t) = (F(t) + F(-t))/2$ is the GF of $\sum_{k=0}^{\infty} a_{2k} t^{2k}$.

**Proof** The proof follows easily from the fact that $F(-t)$ has coefficients with negative sign when $t$ index is odd so that it cancels out when we sum both. Similarly, $G(t) = (F(t) - F(-t))/2 = \sum_{k=0}^{\infty} a_{2k+1} t^{2k+1}$. If $F(t) = \sum_{k=0}^{m} a_k t^k$ where the upper limit is $m$, then $G(t) = \frac{1}{2}(F(t) + F(-t))$ is the GF of $\sum_{k=0}^{\lfloor m/2 \rfloor} a_{2k} t^{2k}$, and $G(t) = \frac{1}{2}(F(t) - F(-t))$ is the OGF of $\sum_{k=0}^{\lfloor m/2 \rfloor} a_{2k+1} t^{2k+1}$. Here $\lfloor x \rfloor$ denotes the greatest integer $\leq x$ (see Chap. 4). If $F(t)$ is an OGF, $(1 + F(t))^n = \sum_{k=0}^{n} \binom{n}{k} F(t)^k$ is also a GF. Arbitrary powers of an OGF can be found recursively using Euler's method (Knuth 1998). $\square$

The Mittag-Lefler OGF is defined as $E(x, \lambda) = \sum_{k=0}^{\infty} x^k / \Gamma(1 + \lambda k)$ (by assuming $a_k = 1/\Gamma(1 + \lambda k)$), and its extension as

$$E(x, \lambda, \mu) = \sum_{k=0}^{\infty} x^k / \Gamma(\mu + \lambda k) \quad \text{where} \quad \lambda > 0, \mu \geq 1. \tag{1.13}$$

It is an extension of exponential function discussed below (as it reduces to the later for $\lambda = \mu = 1$), and finds applications in the solution of fractional order PDE and in physical chemistry.

If an OGF satisfies a simple equation (like a linear or quadric equation along with other simple functions), it is possible to express the coefficients in terms of another equation obtained by solving for the GF. Suppose the OGF satisfies F(t) = 1+t F(t)$^2$. Solving this gives F(t) = $\frac{1}{2t}(1 - \sqrt{1 - 4t})$. The coefficients can now be extracted using an infinite series expansion of the radical.

### 7. OGF of $F_n = F_{n-1} + F_{n-2}$.

Find OGF for the Fibonacci numbers defined as $F_n = F_{n-1} + F_{n-2}$, with $F_0 = 0$ and $F_1 = 1$. Hence obtain the OGF for $(F_0, 0, F_2, 0, F_4, 0, \cdots)$.

Let

$$F(t) = F_0 + F_1 t + F_2 t^2 + \cdots + F_n t^n + \cdots \tag{1.14}$$

be the OGF. Multiply both sides of (1.14) by $t$ and $t^2$ respectively to get

$$t\, F(t) = F_0 t + F_1 t^2 + F_2 t^3 + \cdots + F_n t^{n+1} + \cdots \tag{1.15}$$

$$t^2\, F(t) = F_0 t^2 + F_1 t^3 + F_2 t^4 + \cdots + F_n t^{n+2} + \cdots$$

Subtract from (1.14) to get

$$F(t)[1 - t - t^2] = F_0 + (F_1 - F_0)t + (F_2 - F_1 - F_0)t^2 + \cdots + (F_n - F_{n-1} - F_{n-2})t^n + \cdots \tag{1.16}$$

As $F_n = F_{n-1} + F_{n-2}$, each of the coefficients vanish except the first two. As $F_0 = 0$ and $(F_1 - F_0) = 1 - 0 = 1$, we get $F(t)[1 - t - t^2] = t$, from which $F(t) = t/(1 - t - t^2)$. To get the OGF of $(F_0, 0, F_2, 0, F_4, 0, \cdots)$, use Theorem 1.2 to get $G(t) = [F(t) + F(-t)]/2$. This gives $G(t) = \frac{1}{2}[t/(1 - t - t^2) - t/(1 + t - t^2)]$. The OGF for $k^{th}$ powers of Fibonacci numbers can be found in Riordan (1979).

**Problem 1.6** Find the OGF of $\{F_{2k}\}$ where $F_k$ is the $k^{th}$ Fibonacci number.

**Problem 1.7** Prove that the OGF of $\{F_{m+n}\}$ where $F_k$ is the $k^{th}$ Fibonacci number is $\sum_{k=0}^{\infty} F_{m+k} t^k = (F_m + F_{m-1}t)/(1 - t - t^2)$.

## 1.4    Exponential Generating Functions (EGF)

The primary reason for using a GF in applied scientific problems is that it represents an entire sequence as a single mathematical function in dummy variables, which is easy to investigate and manipulate algebraically. Whereas an OGF is used to count coefficients involving combinations, an EGF is used when the coefficients are permutations or functions thereof. The EGF is preferred to OGF in those applications in which order matters. Examples are counting strings of fixed width formed using binary digits $(0, 1)$. Although bit strings 100 and 010 contain two 0's and one 1, they are distinct.

**Definition 1.2**  The function

$$H(x) = \sum_{n=0}^{\infty} a_n x^n / n! \qquad (1.17)$$

is called the exponential generating function (EGF). There are two ways to interpret the divisor of $n^{th}$ term. Writing it as

$$H(x) = \sum_{n=0}^{\infty} b_n x^n \quad \text{where} \quad b_n = a_n/n! \qquad (1.18)$$

gives an OGF. But it is the usual practice to write it either as (1.17) or as

$$H(x) = \sum_{n=0}^{\infty} a_n e_n x^n \quad \text{where} \quad e_n = 1/n! \qquad (1.19)$$

where we have two independent multipliers. As the coefficients of exponential function $(\exp(x))$ are reciprocals of factorials of natural numbers, they are most suited when terms of a sequence grow very fast. EGF is a convenient way to manipulate such infinite series. Consider the expansion of $e^x$ as

$$e^x = \sum_{n=0}^{\infty} a_n x^n, \qquad (1.20)$$

where $a_n = 1/n!$. This can be considered as the OGF of the sequence $1/n!$, namely the sequence $(1, 1/2!, 1/3!, \ldots)$. If $n!$ is always associated with $x^n$, we could alternately write the above as

$$e^x = \sum_{n=0}^{\infty} a_n x^n / n!, \qquad (1.21)$$

where $a_n = 1$ for all values of $n$.

As the denominator $n!$ grows fast for increasing values of $n$, it can be used to scale down functions that grow too fast. For example, number of permutations, number of subsets of an $n$-element set which has the form $2^n$, etc., when multiplied by $x^n/n!$ will often result in a converging sequence (Bona 2012).[3] Thus, $e^{2x}$ can be considered as the GF of the number of subsets.

These coefficients grow at different rates in practical applications. If this growth is too fast, the multiplier $x^n/n!$ will often ensure that the overall series is convergent. Each $a_n$ is associated with a function of the dummy variable in univariate GFs. The terms $x^n$ in OGF and $x^n/n!$ in EGF are known as "kernels" in analysis of algorithms, integral equations, and many engineering fields. Many new GFs can be obtained by varying the kernel (for example, logarithmic GF has kernel $x^n/n$, and Lambert GF has kernel $x^n/(1-x^n)$).

Quite often, both the OGF and EGF may exist for a sequence. As an example, consider the sequence $1, 2, 4, 8, 16, 32, \ldots$ with $n^{th}$ term $a_n = 2^n$. The OGF is $1/(1-2x)$, whereas the EGF is $e^{2x}$. Thus, the EGF may converge in lot many problems where the OGF either is not convergent or does not exist. Consider $(1+x)^n = \sum_{k=0}^{n} \binom{n}{k} x^k$. We could write this as an EGF by expanding $\binom{n}{k} = n!/[k!(n-k)!]$ as

$$(1+x)^n = \sum_{k=0}^{n} n!/(n-k)! \ (x^k/k!) \tag{1.22}$$

so that $a_k = n!/(n-k)!$ (which happens to be $P(n,k)$, the number of permutations of $n$ things taken $k$ at a time).

When the OGF is of the form $1/(1+ax)$, where $a$ is positive or negative real number, the EGF can be obtained by replacing $x$ by $(-\text{sign}(a)/a)[1-\exp(ax)]$. As shown above, the OGF of $(1, 1, 1, \ldots)$ is $1/(1-x)$. Replace $x$ by $[1-\exp(-x)]$ to get the EGF as $1/(1-[1-\exp(-x)]) = 1/\exp(-x) = \exp(x)$. The EGF could also be obtained by integrating the OGF (Sedgewick and Flajolet 2013), or by Sumudu transforms (Zhang et al. 2021).

Put $a = -1$ and multiply the numerator and denominator by $n!$ to get $\sum_{n=0}^{\infty} n! \times x^n/n! = 1/(1-x)$. This shows that the EGF of $(1, 2!, 3!, \ldots)$ is $1/(1-x)$. As the number of permutations of size $n$ is $n!$, this is the GF for permutations.

Now consider

$$\left(e^{bt} - e^{at}\right)/(b-a) = t/1! + (b^2-a^2)/(b-a)t^2/2! + (b^3-a^3)/(b-a)t^3/3! + \cdots$$

$$+ (b^n - a^n)/(b-a)t^n/n! + \cdots . \tag{1.23}$$

The general solution to the Fibonacci recurrence is of the form $(\beta^n - \alpha^n)/(\beta - \alpha)$ which matches the coefficient of $t^n/n!$ in Eq. (1.23) with "a" replaced by $\alpha$ and "b" replaced by

---

[3] The sequence $\sum_{n\geq 0} n! x^n$ converges only at $x = 0$.

$\beta$ where $\alpha = (1 - \sqrt{5})/2$ and $\beta = (1 + \sqrt{5})/2$. This means that the EGF for Fibonacci numbers is

$$\left(e^{\beta t} - e^{\alpha t}\right)/(\beta - \alpha) = F_0 + F_1 t/1! + F_2 t^2/2! + F_3 t^3/3! + \cdots + F_n t^n/n! + \cdots ,$$
(1.24)

where $F_n$ is the $n^{th}$ Fibonacci number.

Another reason for the popularity of EGF is that they have interesting product and composition formulas. See Chap. 2 for details. The EGF has $x^n/n!$ as multiplier. Some researchers have extended it by replacing $n!$ by the $n^{th}$ Fibonacci number or its factorial $F_n!$ resulting in "Fibonential" GF. Several extensions of EGF are available for specific problems. For example, the Mittag-Lefler GF discussed above can also be considered as an extension of EGF. Another is the tree-like GF (TGF) defined as $F(x) = \sum_{n \geq 0} a_n n^{n-1} x^n/n!$, which has applications in algorithm analysis, coding theory, etc. Nelsen & Schmidt's EGF for the number of preferential arrangements (a partition of a set along with a linear ordering of the blocks) of a finite set S={1,2,$\cdots$,n}, and that of the number of chains in the power set of S is given by $G(k, x) = e^{kx}/(2 - e^x)$, for $k \in N_0$. Nkonkobe and Murali (2017) extended the above to restricted barred preferential arrangements with EGF $G(k, j, x) = e^{kx}/(2 - e^x)^j$.

## 8. OGF of balls in urns

Find the OGF for the number of ways to distribute $n$ distinguishable balls into $m$ distinguishable urns (or boxes) in such a way that none of the urns is empty.

Let $u_{n,m}$ denote the total number of ways. Assume that $i^{th}$ urn has $n_i$ balls so that all urn contents should add up to $n$. That gives $n_1 + n_2 + \cdots + n_m = n$ where each $n_i > 0$. Let $x/1! + x^2/2! + x^3/3! + \cdots$ denote the GF for just one urn (where we have omitted the constant term due to our assumption that the urn is not empty; meaning that there must be at least one ball in it). As there are $m$ such urns, the GF for all of them taken together is $(x/1! + x^2/2! + x^3/3! + \cdots)^m$. The coefficient of $x^n$ is what we are seeking. This is the sum $\sum_{n_1+n_2+\cdots+n_m=n} n!/[n_1!n_2! \cdots n_m!]$. Denote the GF for $u_{n,m}$ by $H(x)$. Then

$$H(x) = \sum_{k=m}^{n} a_k x^k/k! = (x/1! + x^2/2! + x^3/3! + \cdots)^m,$$
(1.25)

where $a_k = u_{k,m}$ and the lower limit is obviously $m$ (because this corresponds to the case where each urn gets one ball), and upper limit is $n$. The RHS is easily seen to be $(e^x - 1)^m$. Expand it using binomial theorem to get

$$(e^x - 1)^m = e^{mx} - \binom{m}{1}e^{(m-1)x} + \binom{m}{2}e^{(m-2)x} - \cdots(-1)^m.$$
(1.26)

Expand each of the terms of the form $e^{kx}$ and collect coefficients of $x^n$ to get the answer as

$$u_{n,m} = m^n - \binom{m}{1}(m-1)^n + \binom{m}{2}(m-2)^n - \cdots (-1)^{m-1}\binom{m}{m-1} \qquad (1.27)$$

because the last term in the above expansion does not have $x^n$.

### 9. EGF of Laguerre polynomials

Prove that the Laguerre polynomials are the EGF of $(-1)^k\binom{n}{k}$ for $k = 0, 1, 2, \ldots, n$.

The Laguerre polynomials are defined as

$$L_n(x) = \sum_{k=0}^{n}(-1)^k\binom{n}{k}/k!x^k. \qquad (1.28)$$

Writing this as $\sum_{k=0}^{n}(-1)^k\binom{n}{k}(x^k/k!)$ shows that the EGF of $(-1)^k\binom{n}{k}$ is the classical Laguerre polynomial.

## 1.5    Pochhammer Generating Functions

Analogous to the formal power series expansions given above, we could also define Pochhammer GF as follows. Here the series remains the same but the dummy variable is either rising factorial or falling factorial type.

### 1.5.1   Rising Pochhammer GF (RPGF)

If $(a_0, a_1, \ldots)$ is a sequence of numbers, the RPGF is defined as

$$\text{RPGF}(x) = \sum_{k \geq 0} a_k x^{(k)}, \qquad (1.29)$$

where $x^{(k)} = x(x+1)\ldots(x+k-1)$, and $x^{(0)} = 1$. Some authors use the notation $< x >_n$ for the rising factorial, and $(x)_n$ for the falling factorial. Note that this is different from the OGF or EGF obtained where the known coefficients follow the Pochhammer factorial as

$$\text{RPGF}(x) = \sum_{k \geq 0} a^{(k)} x^k, \qquad (1.30)$$

where $a^{(k)} = a(a+1)\ldots(a+k-1)$. An example of an EGF of this type is

$$(1-x)^{-b} = \sum_{k=0}^{\infty} b^{(k)} x^k / k! \tag{1.31}$$

which is the EGF for Pochhammer numbers $b^{(k)}$ (see also hypergeometric series given below).

### 1.5.2 Falling Pochhammer GF (FPGF)

The dummy variable is of type falling factorial in this sort of GF.

$$\text{FPGF}(x) = \sum_{k \geq 0} a_k x_{(k)}. \tag{1.32}$$

Consider Stirling number of second kind (discussed below). The FPGF is $x^n$ because $\sum_{k=0}^{n} S(n,k) x_{(k)} = x^n$. The OGF of Stirling number of first kind is $\sum_{k=0}^{n} s(n,k) x^k = x_{(n)}$. The EGF of Stirling number of first kind is $\sum_{n=k}^{\infty} s(n,k) x^n / n! = (1/k!)(\log(1+x))^k$ and that of Stirling number of second kind is $\sum_{n=k}^{\infty} S(n,k) x^n / n! = (e^x - 1)^k / k!$. The falling factorial can also be applied to the constants to get falling EGF for Pochhammer numbers $b^{(k)}$.

$$\text{FPGF} = \sum_{k=0}^{\infty} b_{(k)} x^k / k!. \tag{1.33}$$

This shows that both the sequence terms and dummy variable can be written using Pochhammer symbol, which results in four different types of GFs for OGF and EGF. For instance, the hypergeometric function can be represented as

$$_2F_1(a, b; 1; x) = \sum_{n=0}^{\infty} (a)^{(n)} (b)^{(n)} x^n / (n!)^2, \tag{1.34}$$

and the general hypergeometric function is defined in terms of rising Pochhammer EGF of $a^{(k)} b^{(k)} / c^{(k)}$ as $F(a, b, c; x) = \sum_k \frac{a^{(k)} b^{(k)}}{c^{(k)}} \frac{x^k}{k!}$ where $a^{(k)}$ denotes the rising Pochhammer number. The falling dummy variable technique was introduced by James Stirling in 1725 to represent an analytic function $f(z)$ in terms of difference polynomials as $f(z) = \sum_{k=0}^{\infty} a_k z_{(k)}$. Some examples are $z^2 = z(z-1) + z$, $z^3 = z(z-1)(z-2) + 3z(z-1) + z$, $z^4 = z(z-1)(z-2)(z-3) + 6z(z-1)(z-2) + 7z(z-1) + z$, etc.

### 10. OGF for graph coloring

Find OGF for the number of ways to color the vertices of a complete graph $K_n$.

Fix any of the nodes and give a color to it. There are $n - 1$ neighbors to it so that there are $n - 1$ choices. Continue arguing like this until a single node is left. Hence, the GF is $x(x - 1)(x - 2) \ldots (x - n + 1)$ or $x_{(n)}$ as the required GF.

**Problem 1.8** Prove that the OGF of the sequence $\sum_{n=0}^{\infty} \sum_{k=0}^{n} \binom{n}{k} x^k y^n$ is $1/[1\text{-}(1+x)y]$.

**Problem 1.9** Find the EGF of the sequence defined as $a_n = 1/[(n + 1)(n + 2)]$.

## 1.6   Dirichlet Generating Functions

Dirichlet GF (DGF) of a sequence $\sigma = (a_1, a_2, \cdots)$ is defined as

$$G(\sigma, s) = \sum_{n=1}^{\infty} a_n/n^s = a_1 + a_2/2^s + a_3/3^s + \cdots \tag{1.35}$$

As $k^{-s} = e^{-s \log(k)}$, this could also be written as $\sum_{n=1}^{\infty} a_n \exp(-s \log(n))$. Note that these are not strictly formal power series, and that the dummy variable by convention is chosen as s. The series (1.35) converges for $Re(s) > 1$ and diverges as $s \to 1^+$. If the radius of absolute convergence $R$ is finite, the coefficient $a_n = n^{\sigma} \lim_{T \to \infty} (1/(2T)) \int_{-T}^{T} G(\sigma + 2it) n^{it} dt$ for large $\sigma > R$. For $a_n = n$ we get the DGF as $\sum_{n \geq 1} n/n^s = \sum_{n \geq 1} 1/n^{s-1} = \zeta(s - 1)$. In general, if $a_n = n^m$, the DGF is $\sum_{n \geq 1} n^m/n^s = \sum_{n \geq 1} 1/n^{s-m} = \zeta(s - m)$. A special DGF is $G(\sigma, s) = \sum_{n=0}^{\infty} a_n/(2^n)^s = a_0 + a_1/2^s + a_2/4^s + a_3/8^s + a_4/16^s + \cdots$, which becomes an OGF by the substitution $z = 2^{-s}$ using the fact that $(2^m)^n = 2^{mn}$ in the denominator. Instead of powers of 2 in the denominator, we could also have powers of higher numbers or prime powers $(a_n/(p^n)^s)$ to get DGF that have OGF as analogues. Replace $n$ by $F_n$ in (1.35) to get Fibonacci-Dirichlet GF $\sum_{n=1}^{\infty} a_n/F_n^s$. Likewise, an exponential-Dirichlet GF follows as $\sum_{n=1}^{\infty} a_n/(n!)^s$. When $a_n = F_n$ in (1.35), we get $\sum_{n=1}^{\infty} F_n/n^s$.

The sum of two Dirichlet GFs is of the same type because the denominators are identical. They represent the sum of two interfering probability amplitudes in quantum mechanics.[4] A harmonic Dirichlet GF results when $a_n = H_n$ in (1.35), where $H_n$ is the $n^{th}$ harmonic number. If $a_n$ are multiplicative modulo n (i.e. gcd(m,n) = 1 $\Rightarrow$ $a_m a_n = a_{mn}$ then $G(\{a_n\}, s) = \prod_{p:prime} \sum_{k=0}^{\infty} a_{p^k}/p^{ks}$. Note that the z-transform of a sequence $\{x_k\}$ used in DSP, coding theory and communications engineering is a function of a complex variable z as $Y(z) = \sum_{k=0}^{\infty} x_k/z^k$ where the denominator is $z^k$ instead of $k^z$.

**Problem 1.10** Find an explicit expression for the Fibonacci-Dirichlet OGF $\sum_{k=0}^{\infty} F_{m+k}/k^x$. The Hurwitz GF (HGF) is an extension in which the denominator is modified as

---

[4] Some authors define it in the complex domain as $\sum_{n=1}^{\infty} a_n [\exp(i\theta_n)]^{-s}$.

**Table 1.2** Dirichlet GF

| $a_k$ | Sequence | f(s) |
|---|---|---|
| 1 | ( 1,1,1,1··· ) | $\zeta(s)$ |
| $\mu(n)$ | (1,-1,-1,0,-1,1,-1,0,0,···) | $1/\zeta(s)$ |
| d(n) | (1,2,2,3,2,4,2,4,···) | $[\zeta(s)]^2$ |
| $\phi(n)$ | (1,1,2,2,4,2,6,4,6,4,..) | $\zeta(s-1)/\zeta(s)$ |
| log(k) | (log(2), log(3), log(4), ···) | $-\zeta'(s)$ |
| $\lambda(k)$ | $(\lambda(1), \lambda(2), \lambda(3), \lambda(4), ···)$ | $-\zeta'(s)/\zeta(s)$ |

$\mu(n)$ is the Mobius function

$$H(\sigma, c, s) = \sum_{n=1}^{\infty} a_n/(n+c)^s = a_1/(1+c)^s + a_2/(2+c)^s + a_3/(3+c)^s + \cdots \quad (1.36)$$

When $a_k = 1 \forall k$, the DGF becomes Riemann zeta function $\zeta(s) = \sum_{n=1}^{\infty} n^{-s}$, which has a Euler product representation $\zeta(s) = \prod_p 1/[(1-p^{-s}]$ where p's are primes (ref Table 1.2). It is also related to the Stirling numbers of first kind as $\zeta(k+1) = \sum_n (-1)^{n-k} s(n, k)/[n!n]$. For $s \geq 2$, the zeta function can be represented in terms of Bernoulli numbers as $\zeta(s) = (-1)^{k/2+1}(2\pi)^k/[2 * k!]B_k$ and $\zeta(1-k) = -B_k/k$. A similar alternating series is $\eta(s) = \sum_{n=1}^{\infty}(-1)^{n+1}n^{-s}$, which is related to $\zeta(s)$ as $(1-2^{1-s})\zeta(s) = \eta(s)$ (this is easily derived by adding and subtracting $2(1/2^s + 1/4^s + \cdots)$ and identifying $\zeta(s)$). Dirichlet series, as well as finite linear combinations of them including their analytical continuations to the complex domain are used in quantum physics to represent probability amplitudes for measurements on time-dependent systems (Feiler and Schleich 2013). The linear creep function of concrete at constant temperature and water content can be approximated by Dirichlet series with variable coefficients (Basant and Wu 1973).

**Problem 1.11**  If d(n) is the number of divisors of a positive integer $n > 2$, show that the DGF of the sequence $(d(n))^2$ is $\zeta^4(s)/\zeta(2s)$.

**Problem 1.12**  Prove that the DGF of $\binom{n}{2}$ is $(\zeta(s-2) - \zeta(s-1))/2$. Hint: $\sum_{n=1}^{\infty} \binom{n}{2}/n^2 = (\sum_{n=1}^{\infty} n^2/n^s - \sum_{n=1}^{\infty} n/n^s)/2$.

## 1.7    Other Generating Functions

There are many other GFs used in various fields. Some of them are simple modifications of the dummy variable while others are combinations. It is called characteristic polynomial or weight enumerator in coding theory. If C is a binary code of length n, and $a_k$ is the number of codewords of weight $k$, then $A(z) = \sum_{k=0}^{n} a_k z^k$ is called a weight enumerator of $C$ where $a_0 = 1$. The OGF and EGF can also be defined for sequence of functions. If $\Omega = \{f_1(x), f_2(x), \cdots\}$ are irreducible functions (say polynomials) bounded in a certain interval, the OGF and EGF are respectively $F(x) = \sum_{k\geq 1} f_k(x)x^k$ and $H(x) = \sum_{k\geq 1} f_k(x)x^k/k!$. This type of GF is used in convolutional coding theory where $f_k(x) = I_k(x^n)$ where $I_k()$ are input sequences in which every n-tuple of information bits ends in n-k zeros (van Lint 1999). The trigonometric, inverse-trigonometric and hyperbolic functions (terms multiplied by constants) also can be considered as GFs as they possess infinite series expansions. They may also be combined with polynomials (e.g.: $x^k \cos(kx)$) to get a variety of new GFs.

### 1.7.1   Lambert Generating Function

The Lambert GF (LGF) of a sequence has the form

$$G(t) = \sum_{n=1}^{\infty} a_n t^n/(1 - t^n) \text{ for } -1 < t < 1. \tag{1.37}$$

Expand $(1 - t^n)$ in the denominator as an infinite series and combine with the terms in the numerator to get $\sum_{n=1}^{\infty} a_n t^{mn} = \sum_{k=1}^{\infty} b_k t^k$ where $b_k = \sum_{d|k} a_k$ and $m = 1, 2, 3, \cdots$. An extension of LGF is Euler's GF given by $E(t) = \sum_{n=1}^{\infty} a_n t^n/[(1 - t)(1 - t^2) \cdots (1 - t^n)]$ for $-1 < t < 1$. Put $a_n = F_n$ in (1.37) to get Fibonacci-Lambert GF, and $a_n = H_n$ to get Harmonic-Lambert GF.

### 1.7.2   Exponentio-Ordinary GF

We defined the OGF of a sequence $\sigma = (a_0, a_1, a_2, a_3, \ldots a_n, \ldots)$ as $F(t) = \sum_{n=0}^{\infty} a_n t^n$. Put $t = e^x = \exp(x)$ to get

$$F(e^x) = \sum_{n=0}^{\infty} a_n \exp(x)^n = \sum_{n=0}^{\infty} a_n \exp(nx). \tag{1.38}$$

This is called exponentio-ordinary GF (EOGF). It is used to model radical polymerization at low conversions in polymer chemistry (Chap. 4). If $a'_n$s are probabilities of a discrete distribution (so that $\sum_{n=0}^{\infty} a_n = 1$), it is called moment generating function in statistics ($P_x(e^t) = M_x(t)$; Chap. 3). Similarly, we can put $t = \log(x)$ to get $F(\log(x)) = \sum_{n=0}^{\infty} a_n \log(x)^n$. These substitutions can also be made in the EGF. For example, putting the dummy variable as $\log(x)$ in the EGF gives exponentio-logarithmic GF as $G(\log(x)) = \sum_{n=1}^{\infty} a_n \log(x)^n/n!$.

### 1.7.3 Logarithmic GF

The logarithmic GF (LogGF) is defined as

$$F(x) = \sum_{n=1}^{\infty} a_n x^n / n. \tag{1.39}$$

This can be considered as the OGF of the sequence $\{a_n/n\}$. The LogGF of the sequence $(1, 1, \cdots)$ is $-\log(1-x) = \log(1/(1-x))$. Likewise, the LogGF of the alternating sequence $(1, -1, 1, -1 \cdots)$ is $\log(1+x)$. Add them together to get the LogGF of the sequence $(1, 0, 1, 0 \cdots)$ as $\frac{1}{2}[\log(1+x) - \log(1-x)] = \frac{1}{2}[\log[(1+x)/(1-x)]$, which happens to be the OGF of 1, 1/3, 1/5, 1/7, ... (Page 42, Chap. 2). Similarly, subtraction gives the LogGF of the sequence $(0, 1, 0, 1 \cdots)$ as $-\frac{1}{2}[\log(1-x) + \log(1+x)] = -\frac{1}{2}[\log(1-x^2)]$. The DGF and LogGF coincides when the dummy variable is unity. As in the case of OGF and EGF, we could find convolution and powers of the LogGF as well.

**Problem 1.13** If $F(x)$ is the LogGF of a sequence, find the sequence whose LogGF is $[F(x)]^2$.

### 1.7.4 Special Functions as Generating Function

Special functions of mathematics could also be considered under the framework of GF. Consider the GF

$$J_n(t) = \sum_{k=0}^{\infty} a_k t^{n+2k}, \quad \text{where} \quad a_k = (-1)^k / [k!(n+k)!2^{n+2k}], \tag{1.40}$$

which is the Bessel function of first kind with order n. This can be extended to non-integer values of n by replacing factorial $(n+k)!$ by gamma function $\Gamma(n+k+1)$. Bessel function of integer order can be generated as coefficients of

$$\exp(\frac{x}{2}(t - 1/t)) = \sum_{n=-\infty}^{\infty} t^n J_n(x). \tag{1.41}$$

As the RHS involves negative indices, it is known as Laurent series. Differentiating wrt $x$, it is easy to prove that it satisfies the recurrence $2J_n'(x) = J_{n-1}(x) - J_{n+1}(x)$. Next differentiate wrt the dummy variable and compare like powers of $t$ to get the classical recurrence as $(2n/x)J_n(x) = J_{n-1}(x) + J_{n+1}(x)$. The Legendre polynomials satisfy the recurrence $(n+1)P_{n+1}(t) = (2n+1)t P_n(t) - n P_{n-1}(t)$, for $n \geq 1$, and have the OGF $\sum_{n=0}^{\infty} P_n(t)x^n =$

$1/\sqrt{(1 - 2tx + x^2)}$ for $|t| < 1$. The initial values are $P_0(t) = 1$ and $P_1(t) = t$. Put $t = 1$ to get $\sum_{n=0}^{\infty} P_n(1)x^n = 1/\sqrt{(1 - 2x + x^2)} = 1/(1 - x)$.

### 1.7.5  Auto-covariance Generating Function

Consider the GF

$$P(z) = 2\,a_0 + a_1(z + 1/z) + a_2(z^2 + 1/z^2) + \cdots + a_k(z^k + 1/z^k) + \cdots, \qquad (1.42)$$

where $a_k$'s are the known coefficients. This is used in some of the time series modeling. Separate the $z$ and $1/z$ terms to get $a_0 + a_1 z + a_2 z^2 + \cdots +$ and $a_0 + a_1/z + a_2/z^2 + \cdots +$ which can be written as two separate GF as $G_1(z) + G_2(1/z)$. See Shishebor et al. (2006) for autocovariance GFs for periodically correlated autoregressive processes.

### 11. OGF of Auto-correlation

If the auto-correlation sequence of a discrete time process $X[n]$ is given by $R[n] = a^{|n|}$ where $|a| < 1$, find the auto-correlation GF.

Split the range from $(-\infty$ to $-1)$ and $(0$ to $\infty)$ to get the OGF as $\sum_{n=-\infty}^{-1} a^{-n} t^{-n} + \sum_{n=0}^{\infty} a^n t^n$. This simplifies to $(a/t)(1 - a/t)^{-1} + 1/(1 - at) = a/(t - a) + 1/(1 - at)$.

### 1.7.6  Information Generating Function (IGF)

Let $p_1, p_2, \ldots, p_n$ be the probabilities of a discrete random variable, so that $\sum_{k=1}^{n} p_k = 1$. The IGF of a discrete random variable is defined as $I(t) = -\sum_{k=1}^{n} p_k^t$ where auxiliary variable $t$ is real or complex depending on the nature of the probability distribution. Differentiate w.r.t. $t$ and put $t = 1$ to get

$$H(p) = (\partial/\partial t)\,I(t)|_{t=1} = -\sum_{k=1}^{n} p_k \log(p_k) \qquad (1.43)$$

which is Shannon's entropy for log to the base 2. Differentiate (1.43) $m$ times and put $t = 1$ to get

$$H(p)^{(m)} = (\partial/\partial t)^m I(t)|_{t=1} = -\sum_{k=1}^{n} p_k (\log(p_k))^m \qquad (1.44)$$

which is the expected value of $m^{th}$ power of $\log(p)$.

**Problem 1.14** Find the OGF and EGF of the sequence $a_n = 1/[(n + 1)(n + 2)]$.

**Problem 1.15**  Find the OGF of the sequence $a_n = (-1)^{(n-1)/2}/n!$ for n an odd integer, and $a_n = 0$ otherwise.

## 1.8  Generating Functions in Number Theory

There are many useful GFs used in number theory. Some of them are discussed in subsequent chapters. Examples are Fibonacci and Lucas numbers, Bell numbers, Catalan numbers, Lah numbers, Bernoulli numbers, Stirling numbers, Euler's $\sigma(n)$, etc.

### 1.8.1  Lah Numbers

The Lah numbers introduced by Ivo Lah (1896–1979) in 1955 are defined as $L(n, k) = \binom{n-1}{k-1}n!/k! = (n-1)!n!/[(k-1)!k!(n-k)!] = \Gamma(n)\Gamma(n+1)/[\Gamma(k)\Gamma(k+1)\Gamma(n-k+1)]$ for $1 \leq k \leq n$. Signless Lah numbers can be used to express the rising Pochhammer factorial powers in terms of falling powers, and *vice versa*.

$$x^{(n)} = \sum_{k=0}^{n} L(n, k)x_{(k)} \text{ for } n \geq k \geq 1. \tag{1.45}$$

Similarly, we could write $x(x-1)\cdots(x-n+1) = \sum_{k=1}^{n}(-1)^{n-k}L(n,k)x(x+1)\cdots(x+k-1)$ and $x(x+1)\cdots(x+n-1) = \sum_{k=1}^{n}L(n,k)x(x-1)\cdots(x-k+1)$. This is known as "Lah identity". Note that $x^{(n)} = (-1)^n(-x)_{(n)}$. This allows us to write $(-x)_{(n)} = \sum_{k=0}^{n} L(n,k)x_{(k)}$ or equivalently $(x)_{(n)} = |\sum_{k=1}^{n} L(n,k)(-x)_{(k)}|$ (Riordan 1979). The unsigned version represents the number of ways to partition a set of cardinality n> 1 into k non-empty linearly ordered subsets. They satisfy the recurrence $L(n, k) = L(n-1, k-1) + (n+k-1)L(n-1, k)$. It has EGF $\sum_{n=0}^{\infty} L(n, k)t^n/n! = (1/k!)(t/(1-t))^k$. Replacing $t$ by $-t$, this could also be written as $(-1)^k(1/k!)(t/(1+t))^k = \sum_{n=0}^{\infty}(-1)^n L(n, k)t^n/n!$. It is a convolution of Stirling numbers as $L(n, k) = \sum_{j=k}^{n} s(n, j)S(j, k) = \sum_{j=k}^{n} \genfrac{[}{]}{0pt}{}{n}{j}\genfrac{\{}{\}}{0pt}{}{j}{k}$. See Daboul (2013), Boyadzhiev (2016) for a connection among Lah numbers and the n-th derivative of $\exp(1/x)$, and Ghosal et al. (2020) for an application of Lah transform for security and privacy of data in telecommunication engineering.

### 1.8.2  Rook Polynomial Generating Function

Consider an $n \times n$ chessboard. A rook can be placed in any cell such that no two rooks appear in any row or column. Let $r_k$ denote the number of ways to place $k$ rooks on a chessboard.

Obviously $r_0 = 1$ and $r_1 = n^2$ (because it can be placed anywhere), and $r_n = n!$. The rook polynomial GF can now be defined as

$$R(t) = r_0 + r_1 t + r_2 t^2 + \cdots . \tag{1.46}$$

The rook polynomial for an $n \times n$ chessboard is given by $\sum_{k=0}^{n} k! \binom{n}{k}^2 t^k$. There are $r_1 = n^2$ ways to place one rook. Suppose there are $k$ distinguishable rooks. The first can be placed in $r_1 = n^2$ ways, which marks-off one row and one column (because rooks attack horizontally or vertically). There are $(n-1)^2$ squares remaining so that the second one can be placed in $(n-1)^2$ ways and so on. As the rooks can be arranged among themselves in $k!$ ways, we get the GF as $\sum_{k=0}^{n} n^2(n-1)^2(n-2)^2 \ldots (n-k+1)^2 / k! t^k$. Multiply and divide by $k!$ and use $n(n-1)(n-2) \ldots (n-k+1)/k! = \binom{n}{k}$ to get the result.

### 1.8.3  Stirling Numbers

There are two types of Stirling numbers—the first kind denoted by $s(n,k)$ and the second kind denoted by $S(n,k)$ or $\genfrac{[}{]}{0pt}{}{n}{k}, \genfrac{\{}{\}}{0pt}{}{n}{k}$. Consider a set $S$ with $n$ elements. We wish to partition it into $k(< n)$ nonempty subsets. The $S(n,k)$ denotes the number of partitions of a finite set of size $n$ into $k$ subsets. It can also be considered as the number of ways to distribute $n$ distinguishable balls into $k$ indistinguishable cells with no cell empty. They obey the orthogonal relationship $\sum_{k=m}^{n} (-1)^{k+m} \genfrac{[}{]}{0pt}{}{n}{k} \genfrac{\{}{\}}{0pt}{}{k}{m} = \delta_{nm}$

The Stirling number of second kind satisfies the recurrence relation $S(n,k) = kS(n-1,k) + S(n-1,k-1)$. The OGF for the Stirling numbers of the second kind is given by

$$\sum_{n=0}^{\infty} S(n,k)x^n = x^k / [(1-x)(1-2x)(1-3x) \ldots (1-kx)]. \tag{1.47}$$

A simplified form results when $n$ is represented as a function of $k$ (displaced by another integer):

$$\sum_{n=0}^{\infty} S(m+n,m)x^n = 1/[(1-x)(1-2x)(1-3x) \ldots (1-mx)]. \tag{1.48}$$

## 1.9    Summary

This chapter introduced the basic concepts in GFs. Topics covered include structure of GFs, types of GFs, ordinary and exponential GFs, etc. The Pochhammer GF for rising and falling factorials are also introduced. Some special GFs like auto-covariance GFs, information GFs, and those encountered in number theory and graph theory are briefly described. Multiplying an OGF by $1/(1-x)$ results in the OGF of the partial sums of coefficients. This result is used extensively in Chap. 3.

## References

Basant, Z. P., & Wu, S. T. (1973). Dirichlet series creep function for aging concrete. *Journal of the Engineering Mechanics Division, 99*(2) (1973). https://doi.org/10.1061/JMCEA3.0001741.

Bona, M. (2012). *Combinatorics of permutations* (2nd edn.) CRC Press.

Boyadzhiev, K. N. (2016). Lah numbers, Laguerre polynomials of order negative one, and the n-th derivative of exp(1/x). *Acta University Sapientiae, Mathematica, 8*(1), 22–31. https://doi.org/10.1515/ausm-2016-0002.

Chattamvelli, R., & Jones, M. C. (1995). Recurrence relations for noncentral density, distribution functions, and inverse moments. *Journal of Statistical Computation and Simulation, 52*(3), 289–299. https://doi.org/10.1080/00949659508811679.

Daboul, S., et al. (2013). The Lah numbers and the n-th derivative of exp($1/x$). *Mathematics Magazine, 86*(1), 39–47.

Feiler, C., & Schleich, W. P. (2015). Dirichlet series as interfering probability amplitudes for quantum measurements. *New Journal of Physics, 17*, 063040. https://iopscience.iop.org/article/10.1088/1367-2630/17/6/063040.

Ghosal, S. K., Mukhopadhyay, S., Hossain, S., & Sarkar, R. (2020). Application of Lah transform for security and privacy of data through information hiding in telecommunication. In *Transactions on Emerging Telecommunications Technologies* (pp. 1–20). Wiley. https://doi.org/10.1002/ett.3984.

Hartleb, D., Ahrens, A., Purvinis, O., & Zascerinska, J. (2020). Analysis of free time intervals between buyers at cash register using generating functions. *Proceedings of the 10th international conference on pervasive and parallel computing: Communication and sensors (PECCS2020)* (pp. 42–49). ISBN: 978-989-758-4770.

Knuth, D. E. (1998). *Fundamental algorithms* (Vol. 2). Boston: Addison Wesley.

Lah, I. (1955). Eine neue art von zahlen, hire eigenschaftern and anwendung in Der mathematischen statistic. *Mitteilungsblatt Mathematics Status, 7*, 203–216.

Nkonkobe, S., & Murali, V. (2017). A study of a family of generating functions of Nelsen-Schmidt type and some identities on restricted barred preferential arrangements. *Discrete Mathematics, 340*(5), 1122–1128. https://doi.org/10.1016/j.disc.2016.11.010.

Riordan, J. (1979). *Combinatorial identities*. New York: Wiley.

Sedgewick, R., & Flajolet, P. (2013). *An introduction to the analysis of algorithms*. MA: Addison-Wesley.

Shishebor, Z., Nematollahi, A. R., & Soltani1, A. R. (2006). On covariance generating functions and spectral densities of periodically correlated autoregressive processes. *Journal of Applied Mathematics and Stochastic Analysis, 2006*, Article ID 94746. https://www.hindawi.com/journals/ijsa/2006/094746/ref/.

van Lint, J. H. (1999). *Introduction to coding theory*. Springer.

Zhang, J., Fan, R, & Shen, F. (2021). New method for the computation of generating functions with applications. In *2021 international conference on computational science and computational intelligence (CSCI)*. Las Vegas, NV, USA: IEEE. https://doi.org/10.1109/CSCI54926.2021.00156.

# Operations on Generating Functions

**2**

This chapter discusses common operations on generating functions (GFs) like addition, linear combination, left and right shifting, differentiation and integration, tail-sum etc. More advanced operations like convolution, powers, higher-order derivatives, index-multiply, functions of dummy variables are also briefly discussed. These are applied to both OGF and EGF.

## 2.1 Basic Operations

The first chapter explored several simple sequences. As a sequence is indexed by natural numbers, the best way to express an arbitrary term of a sequence seems to be a closed form as a function of the index $(n)$. For example, $a_n = 2^n - 1$ expresses the $n$th term as a function of $n$. Some of the series encountered in practice cannot be described succinctly as shown above. A recurrence relation connecting successive terms may give us good insights for some sequences. Such a recurrence suffices in most of the problems, especially those used in updating the terms. But it may be impractical to use a recurrence to calculate the coefficients for very large indices (say 10 millionth term). Given a recurrence for two sequences $a_n$ and $b_n$, how do we get a recurrence for $a_n + b_n$ or for $c * a_n$?. This is where the merit of introducing a GF becomes apparent. It allows us to derive new GFs from existing or already known ones.

An infinite sum is denoted by either explicit enumeration of the range as $\sum_{k=0}^{\infty} p_k t^k$, or as $\sum_{k \geq 0} p_k t^k$ where the upper limit is to be interpreted appropriately. Most of the following results are apparent when the coefficient of the power series are represented as sets. For example, if $(a_0, a_1, \ldots)$ and $(b_0, b_1, \ldots)$ are, respectively, the coefficients of $A(t)$ and $B(t)$, then $(a_0 \pm b_0, a_1 \pm b_1, \ldots)$ are the coefficients of the sum and difference $A(t) + B(t)$ and $A(t) - B(t)$; while $(c_0, c_1, \ldots)$ are the coefficients of $A(t) * B(t)$, where $c_0 = a_0 b_0$, $c_1 = a_0 b_1 + a_1 b_0$, etc., $c_n = a_0 b_n + a_1 b_{n-1} + \cdots + a_n b_0$. Two GFs, say $F(t)$ and $G(t)$ are

© The Author(s), under exclusive license to Springer Nature Switzerland AG 2023
R. Chattamvelli and R. Shanmugam, *Generating Functions in Engineering and the Applied Sciences*, Synthesis Lectures on Engineering, Science, and Technology,
https://doi.org/10.1007/978-3-031-21143-0_2

exactly identical when each of the coefficients of $t^k$ are equal for all $k$. This holds good for OGF, EGF, and other aforementioned GFs.

Let $F(t)$ and $G(t)$ defined as

$$F(t) = a_0 + a_1 t + a_2 t^2 + \cdots = \sum_{k=0}^{\infty} a_k t^k \tag{2.1}$$

$$G(t) = b_0 + b_1 t + b_2 t^2 + \cdots = \sum_{k=0}^{\infty} b_k t^k \tag{2.2}$$

be two arbitrary OGF and with compatible coefficients. Here, $a_k$ (respectively $b_k$) is the coefficient of $t^k$ in the power series expansion of $F(t)$ (resp. $G(t)$).

### 2.1.1  Extracting Coefficients

Quite often, the GF approach allows us to find a compact representation for a sequence at hand, and to extract the terms of a sequence from its GF using simple techniques (power series expansion, differentiation, etc.). Thus it is a two-way tool. A general method to extract the coefficient is using Cauchy's contour integral. However, the coefficients can often be extracted much more easily from a GF. There are situations where we may have to use other mathematical techniques like partial fractions, differentiation, logarithmic transformations, synthetic division etc. Suppose a GF is in the form $p(x)/q(x)$ where $p(x)$ is either a polynomial or a constant. It is said to be in *normal form* if there are no common factors. If the degree of $p(x)$ is less than that of $q(x)$, and $q(x)$ contains products of functions, we may split the entire expression using partial fractions to simplify the coefficient extraction. If the degree of $p(x)$ is greater than or equal to that of $q(x)$, we may use synthetic division to get an expression of the form $r(x) + p1(x)/q(x)$, where the degree of $p1(x)$ is less than that of $q(x)$, and proceed with partial fraction decomposition of $p1(x)/q(x)$. A rational function $p(x)/q(x)$ is said to be in standard form (normalised) if they are devoid of common factors. The algebraic roots of $q(x)$ are called 'poles', and their reciprocals called 'roots'. On occasion, the denominator expression $q(x)$ will have complex roots. However, when expanded as an infinite series, the complex part will cancel out giving a real constant. Alternatively, multiply both numerator and denominator by a suitable expression to get a simpler denominator. For instance, if the denominator is $(1 + x + x^2)$ multiply by $(1 - x)$ to get $(1 - x^3)$, which can be expanded as an infinite series. These are illustrated in examples below.

### 2.1.2  Addition and Subtraction

The sum and difference of $F(t)$ and $G(t)$ are defined as

$$F(t) \pm G(t) = (a_0 \pm b_0) + (a_1 \pm b_1)t + (a_2 \pm b_2)t^2 + \cdots = \sum_{k=0}^{\infty} (a_k \pm b_k)t^k. \tag{2.3}$$

Thus, the above result can be extended to any number of GFs as $F(t) \pm G(t) \pm H(t) \pm \cdots$, where each of them are either OGF or EGF. Hence, $F(t) - G(t) + H(t)$ is a valid OGF if each of them is an OGF. This is a special case of the linear combination discussed below. Some authors denote this compactly as $(F - G + H)(t)$. This has applications in several fields like statistics where we work with sums of independently and identically distributed (IID) random variables.

### 2.1.3    Multiplication by Non-Zero Constant

This is called a scalar product or scaling, and is defined as $cF(t) = c \sum_{k=0}^{\infty} a_k t^k = \sum_{k=0}^{\infty} c * a_k t^k = \sum_{k=0}^{\infty} b_k t^k$ where $b_k = c\, a_k$, and $c$ is a non-zero constant. This technique allows us to represent a variety of sequences using a constant times a single GF. For instance, if there are multiple geometric series that differ only in the common ratio, we could represent all of them using this technique. As constant $c$ is arbitrary, we could also *divide* a GF (OGF, EGF, Pochhammer GF, DGF, etc.) by a non-zero constant.

### 2.1.4    Linear Combination

Scalar multiplication can be extended to a linear combination of OGF as follows. Let $c$ and $d$ be two non-zero constants. Then $cF(t) + dG(t) = \sum_{k=0}^{\infty} (ca_k + db_k)t^k$ (where the missing operator is $*$). The constants can be positive or negative, which results in a variety of new GFs. This also can be extended to any number ($n \geq 2$) of GFs. As a special case, the equality of two GFs mentioned in the first paragraph can be proved as follows. Suppose $F(t)$ and $G(t)$ are two GFs. Assume for simplicity that both are OGF. Then by taking $c = +1$ and $d = -1$, we have $F(t) - G(t) = \sum_{k=0}^{\infty} a_k t^k - \sum_{k=0}^{\infty} b_k t^k$, or equivalently $\sum_{k=0}^{\infty} (a_k - b_k)t^k$. Equating to zero shows that this is true only when each of the $a_k'$s and $b_k'$s are equal, proving the result. The same argument holds for EGF, Pochhammer GF, and other GFs.

#### 1. OGF of Arithmetic Progression

Find the OGF of a sequence $(a, a + d, a + 2d, \ldots, a + (n - 1)d, \ldots)$ in arithmetic progression (AP) with initial term '$a$' and common difference $d$.

Write the OGF as

$$F(t) = a + (a + d)t + (a + 2d)t^2 + \cdots + (a + nd)t^n + \cdots . \qquad (2.4)$$

This can be split into two separate OGFs as

$$F(t) = a\left[1 + t + t^2 + \cdots\right] + dt\left[1 + 2t + 3t^2 + \cdots + nt^{n-1} + \cdots\right]. \qquad (2.5)$$

This gives $F(t) = a/(1 - t) + dt/(1 - t)^2 = [a + (d - a)t]/(1 - t)^2$.

### 2.1.5   Shifting

If we have a sequence $(a_0, a_1, \ldots)$, shifting it right by one position means to move everything to the right by one unit and fill the vacant slot on the left by a zero. This results in the sequence $(0, a_0, a_1, \ldots)$, which corresponds to $b_0 = 0$, $b_n = a_{n-1}$ for $n \geq 1$. Similarly, shifting left means to move everything to the left by one position and discarding the very first term on the left, which corresponds to $b_n = a_{n+1}$ for $n \geq 0$. For finite sequence, we may have to fill the vacant terms on the right-most position by zeros. Let $F(t) = a_0 + a_1 t + a_2 t^2 + \cdots$ be the OGF for $(a_0, a_1, \ldots)$. Then $tF(t)$ is an OGF for $(0, a_0, a_1, \ldots)$. In general $t^m F(t) = a_0 t^m + a_1 t^{m+1} + a_2 t^{m+2} + \cdots$. This is the OGF of $(\overbrace{0, 0, \ldots 0}, a_0, a_1, a_2, \ldots)$ where the "zero block" in the beginning has $m$ zeros. Simple algebraic operations can be performed to show that

$$H(t) = \left( F(t) - a_0 - a_1 t - a_2 t^2 - \cdots - a_{m-1} t^{m-1} \right) / t^m = a_m + a_{m+1} t + a_{m+2} t^2 + \cdots$$
(2.6)

which is a shift-left by m positions. In this case $[t^n] H(t) = [t^{m+n}] F(t)$. Shifting is applicable for both OGF and EGF, and will be used in the following paragraphs.

**2. Sequence whose OGF is $(1 + t) G(t)$**

Let $G(t)$ be the OGF of a sequence $(a_0, a_1, \ldots)$. What is the sequence whose OGF is $(1 + t) G(t)$?

Write $(1 + t) G(t) = G(t) + t G(t)$. It follows easily that this is the OGF of $(a_0, a_0 + a_1, a_1 + a_2, \ldots)$ because $t G(t)$ is the aforementioned right-shifted sequence.

**3. OGF of $(1, 3, 7, 15, 31, \ldots)$**

Find the OGF of the sequence $(1, 3, 7, 15, 31, \ldots)$.

The $k$th term is obviously $(2^k - 1)$ for $k \geq 1$. Let $F(t)$ denote the OGF of the given sequence. Then $F(t) = \sum_{k=1}^{\infty} (2^k - 1) t^k$. Split this into two series. The first one is $\sum_{k=1}^{\infty} 2^k t^k = 2t + 4t^2 + \cdots = (1 - 2t)^{-1} - 1$. The second one is $-\sum_{k=1}^{\infty} t^k = -t/(1 - t)$. Hence, $F(t) = 1/(1 - 2t) - t/(1 - t) - 1$, which simplifies to $(1 - 2t + 2t^2)/[(1 - t)(1 - 2t)] - 1$.

**4. OGF of $(1^2, 3^2, 5^2, 7^2, \ldots)$**

Find the OGF and EGF of the sequence $(1^2, 3^2, 5^2, 7^2, \ldots)$.

Let $F(t)$ denote the OGF of the given sequence. The $k$th term is obviously $(2k + 1)^2$ for $k \geq 0$. Hence, $F(t) = \sum_{k=0}^{\infty} (2k + 1)^2 t^k$. Expand $(2k + 1)^2 = 4k^2 + 4k + 1$ and split into three terms to get $F(t) = 4 \sum_{k=0}^{\infty} k^2 t^k + 4 \sum_{k=0}^{\infty} k t^k + \sum_{k=0}^{\infty} t^k$. All of these have already been encountered in previous chapter. Thus, we get $F(t) = 4t(1 + t)/(1 - t)^3 + 4/(1 - t)^2 + 1/(1 - t)$.

Next let $H(t)$ denote the EGF of the given sequence. Then $H(t) = \sum_{k=0}^{\infty}(2k+1)^2 t^k/k!$. Expand $(2k+1)^2 = 4k^2 + 4k + 1$ and split into three terms to get $H(t) = 4\sum_{k=0}^{\infty} k^2 t^k/k! + 4\sum_{k=0}^{\infty} kt^k/k! + \sum_{k=0}^{\infty} t^k/k!$. Write $k^2 = k(k-1) + k$ and split the first term into two EGF's. Then combine the first and second series. This results in

$$H(t) = 4t^2 \sum_{k=2}^{\infty} t^{k-2}/(k-2)! + 8t \sum_{k=1}^{\infty} t^{k-1}/(k-1)! + \sum_{k=0}^{\infty} t^k/k!. \qquad (2.7)$$

This simplifies to $H(t) = (4t^2 + 8t + 1)e^t$.

### 2.1.6 Functions of Dummy Variable

The dummy variable in a GF can be assumed to be continuous in an interval (i.e. in its RoC). Any type of transformation can then be applied on them within that region. A simple case is the scaling of $t$ as $ct$ in an OGF

$$F(ct) = \sum_{k \geq 0} a_k(ct)^k = \sum_{k \geq 0} (c^k a_k)t^k, \qquad (2.8)$$

which is an OGF of $\{c^k a_k\}$ where $c$ can be any real number or fractions of the form $p/q$. By taking $c = 1/2$, we get the series $\{a_k/2^k\}$.

Put $v = 1/t$ to get a reciprocal dummy variable GF(RDV-GF) as $F(v) = a_0 + a_1/v + a_2/v^2 + \cdots$. If $f(t) = \sum_{n=0}^{\infty} a_n t^n$ is the OGF of an entire function, its Borel transform is given by $F(s) = \sum_{n=0}^{\infty} n! \, a_n/s^{n+1}$. Write $1/s = t$ to get $F(1/t)$ as an OGF with coefficient $(n-1)! \, a_{n-1} = \Gamma(n) \, a_{n-1}$. The virial equation used in thermodynamics is a power series expansion of compressibility factor in density (reciprocal of volume) in which $a_k$ are called virial coefficients (and $a_0 = 1$) that depend only on temperature. This also is of type RDV-GF. It also finds applications in Automata theory (Chouraqui 2019). Transforming the dummy variable as $y = (x-1)/x$ results in an OGF $F(x) = \sum_{n=0}^{\infty} a_n \left(\frac{x-1}{x}\right)^n$. This OGF is divergent as $x \to 0$. For $a_n = 1/n$ this converges to $\log_e(x) = \ln(x)$.

**5. OGF of $a_n = 2n + 1$ for $n$ odd, $a_n = n + 1$ otherwise**

Find the OGF of a sequence defined as $a_n = 2n + 1$ if $n$ is odd, and $a_n = n + 1$ if $n$ is even.

The sequence is easily seen to be $\{1, 3, 3, 7, 5, 11, 7, 15, 9, 19, \cdots\}$. First consider the odd terms $\{1, 3, 5, 7, 9, \cdots\}$. It was shown in Page 10 of Chap. 1 that the OGF of odd positive integers is $(1+t)/(1-t)^2$. Replace $t$ by $t^2$ to get the OGF of $\{1, 0, 3, 0, 5, 0, \cdots\}$ as $(1+t^2)/(1-t^2)^2$. Next consider $\{3, 7, 11, 15, \cdots\}$. If 3 is subtracted from each term, we get $\{0, 4, 8, 12, \cdots\}$, which has OGF $4t^3 + 8t^5 + 12t^7 + \cdots$. Take $4t^3$ as common factor and simplify to get $4t^3/(1-t^2)^2$. As we have subtracted 3 from each coefficient, the resulting (subtracted) expression is $-3[t + t^3 + t^5 + t^7 + \cdots]$. Take $t$ outside and simplify to get $-3t/(1-t^2)$. Hence the OGF of $\{3, 7, 11, 15, \cdots\}$ is $4t^3/(1-t^2)^2 - 3t/(1-t^2)$.

Combine with the OGF of $\{1, 0, 3, 0, 5, 0, \cdots\}$ to get $4t^3/(1-t^2)^2 - 3t/(1-t^2) + (1 + t^2)/(1-t^2)^2$. Write $(1+t^2)/(1-t^2)^2 = 1/(1-t^2) + 2t^2/(1-t^2)^2$ to get a simplified expression as $F(t) = (1 - 3t)/(1 - t^2) + 2t^2(1 + 2t)/(1 - t^2)^2$.

### 6. OGF of $a_n = 2n - 1$ for $n$ odd, $a_n = n/2$ otherwise

Find the OGF if $a_n = 2n - 1$ for $n$ odd and $a_n = n/2$ for $n$ even.

The sequence is easily seen to be $\{0, 1, 1, 5, 2, 9, 3, 14, 4 \cdots\}$. The OGF of the even terms is obviously $1/(1 - t^2)$. As done above, subtract 1 from odd coefficients to get the OGF as $4t^3/(1 - t^2)^2$, and the subtracted OGF as $t^3/(1 - t^2)$. Combine with the above to get the OGF as $F(t) = (1 + t^3)/(1 - t^2) + 4t^3/(1 - t^2)^2$.

### 7. OGF of $a_n = ((n + 1)/2^n)$.

Find the OGF of the sequence whose $n$th term is $((n + 1)/2^n)$ for $n \geq 0$.

Let $F(t)$ denote the OGF

$$F(t) = \sum_{n \geq 0} ((n + 1)/2^n)\, t^n = \sum_{n \geq 0} (n + 1)(t/2)^n. \tag{2.9}$$

Putting $s = t/2$ shows that this is the OGF of $(1 - s)^{-2}$. Thus, the OGF of original sequence is $(1 - t/2)^{-2}$ or $1/(1 - t/2)^2$.

## 2.1.7 Convolutions and Powers

The literal meaning of convolution is "a thing, a form or shape that is folded in curved twist or tortuous windings that is complex or difficult to follow." A convolution in GFs is defined for two sequences (usually different or time-lagged as in DSP) with the same number of elements as the sum of the products of terms taken in opposite directions. As an example, the convolution of $(1, 2, 3)$ and $(4, 5, 6)$ is $1 * 6 + 2 * 5 + 3 * 4 = 28$. Consider the product of two GFs

$$\sum_{k=0}^{\infty} a_k t^k * \sum_{k=0}^{\infty} b_k t^k = \sum_{k=0}^{\infty} c_k t^k \quad \text{where} \quad c_k = \sum_{j=0}^{k} a_j b_{k-j}. \tag{2.10}$$

This is called the convolution, which can also be written as $c_k = \sum_{i+j=k} a_j b_i$. As the order of the OGF can be exchanged, it is the same as $\sum_{j=0}^{k} a_{k-j} b_j$. The coefficients of a convolution all lie on a diagonal which is perpendicular to the main diagonal (Table 2.1). This shows that a convolution of two sequences with compatible coefficients can be represented as the product of their GFs. This has several applications in selection problems, signal processing, and statistics. This can be expressed as

**Table 2.1** Convolution as diagonal sum

| ↗ | $b_0$ | $b_1 x$ | $b_2 x^2$ | $b_3 x^3$ | $b_4 x^4$ | ... |
|---|---|---|---|---|---|---|
| $a_0$ | $a_0 b_0$ | $a_0 b_1 x$ | $a_0 b_2 x^2$ | $a_0 b_3 x^3$ | $a_0 b_4 x^4$ | ... |
| $a_1 x$ | $a_1 b_0 x$ | $a_1 b_1 x^2$ | $a_1 b_2 x^3$ | $a_1 b_3 x^4$ | $a_1 b_4 x^5$ | ... |
| $a_2 x^2$ | $a_2 b_0 x^2$ | $a_2 b_1 x^3$ | $a_2 b_2 x^4$ | $a_2 b_3 x^5$ | $a_2 b_4 x^6$ | ... |
| $a_3 x^3$ | $a_3 b_0 x^3$ | $a_3 b_1 x^4$ | $a_3 b_2 x^5$ | $a_3 b_3 x^6$ | $a_3 b_4 x^7$ | ... |
| $a_4 x^4$ | ... | ... | ... | ... | ... | ... |

$$F(t)G(t) = \sum_{k \geq 0} (a_0 b_k + a_1 b_{k-1} + \cdots + a_k b_0)\, t^k. \qquad (2.11)$$

Suppose there are two disjoint sets $A$ and $B$. If the GF for selecting items from $A$ is $A(t)$ and from $B$ is $B(t)$, the GF for selecting from the union of $A$ and $B (A \cup B)$ is $A(t)B(t)$. This can be applied any number of times. Thus, if $A(t)$, $B(t)$, and $C(t)$ are three OGFs with compatible coefficients, the OGF of $A(t)B(t)C(t)$ has $m$th coefficient $u_m = \sum_{i+j+k=m} a_i b_j c_k$.

### 8. OGF of a product

Find the OGF of the product of $(1, 2, 3, 4, \ldots)$ and $(1, 2, 4, 8, 16, \ldots)$.

The first OGF is obviously $1/(1-t)^2$ and second one is $1/(1-2t)$. The convolution is the product of the OGF $1/[(1-t)^2(1-2t)]$. Use partial fractions to break this as $A/(1-t) + B/(1-t)^2 + C/(1-2t)$. Take $(1-t)(1-t)^2(1-2t)$ as a common denominator and equate numerators on LHS and RHS to get $A(1-t)(1-2t) + B(1-2t) + C(1-t)^2 = 1$. Now equate constant, coefficient of $t$, and $t^2$ to get 3 equations $A + B + C = 1$, $3A + 2B + 2C = 0$, $2A + C = 0$. Multiply the first equation by 2 and subtract from the second one to get $A = -2$, from which $C = 4$ and $B = -1$. Thus, the OGF of the product is $4/(1-2t) - 2/(1-t) - 1/(1-t)^2$ with RoC $0 < t < 1/2$.

Now consider two EGFs $H_1(t) = \sum_{k=0}^{\infty} a_k t^k/k!$ and $H_2(t) = \sum_{k=0}^{\infty} b_k t^k/k!$. Their product $H_1(t)H_2(t)$ has "exponential convolution"

$$\sum_{k=0}^{\infty} a_k t^k/k! * \sum_{k=0}^{\infty} b_k t^k/k! = \sum_{k=0}^{\infty} c_k t^k/k! \qquad (2.12)$$

$$c_k = \sum_{j=0}^{k} a_j b_{k-j} k!/(j!(k-j)!) = \sum_{j=0}^{k} \binom{k}{j} a_j b_{k-j}. \qquad (2.13)$$

As the order of the EGF can be exchanged, it is the same as $c_k = \sum_{j=0}^{k} \binom{k}{j} a_{k-j} b_j$.

If $G(s) = \sum_{n=1}^{\infty} a_n/n^s$ and $H(s) = \sum_{n=1}^{\infty} b_n/n^s$ are two DGFs, their convolution is given by $G(s) * H(s) = \sum_{n=1}^{\infty} c_n/n^s$ where $c_n = \sum_{j*k=n} a_j b_k = \sum_{d|n} a_d b_{n/d}$, which are indexed over all positive divisors of n. The first four coefficients are $c_1 = a_1 b_1$,

$c_2 = a_1 b_2 + a_2 b_1$, $c_3 = a_1 b_3 + a_3 b_1$, and $c_4 = a_1 b_4 + a_2 b_2 + a_4 b_1$. These coefficients have a simple form for prime indices. Thus $c_7 = a_1 b_7 + a_7 b_1$. A special case is powers of DGF $[G(s)]^k = (\sum_{n=1}^{\infty} a_n/n^s)^k$, the coefficients of which can be written as $b_n = \sum_{n_1*n_2*\cdots*n_k=n} a_{n_1} a_{n_2} \cdots a_{n_k}$. As in the case of OGF and EGF, the coefficients in a DGF can be multiplicative number-theoretic functions (Wilf 1994). As the derivative of $a^x$ is $a^x \log(a)$, we have $G'(s) = -\sum_{n=1}^{\infty} a_n \log(n)/n^s$. In particular, $\zeta'(s) = -\sum_{n=1}^{\infty} \log(n)/n^s$. Particular cases of DGF are given in Chap. 1.

As convolution is commutative and associative, we could write $F(t)G(t) = G(t)F(t)$ and $F(t)G(t)H(t) = G(t)H(t)F(t) = H(t)F(t)G(t)$, etc. Similarly, product distributes over addition and subtraction. This means that $F(t)[G(t) \pm H(t)] = F(t)G(t) \pm F(t)H(t)$. Powers of an OGF are repeated convolutions. Thus, if $F(t) = \sum_{k=0}^{\infty} a_k t^k$ then $F(t) * F(t) = F(t)^2 = \sum_{k=0}^{\infty} c_k t^k$ where $c_k = \sum_{j=0}^{k} a_j a_{k-j}$, which can also be expressed as

$$[F(t)]^2 = \sum_{k \geq 0} (a_0 a_k + a_1 a_{k-1} + \cdots + a_k a_0) t^k. \tag{2.14}$$

As an illustration of powers of OGF, consider the number of binary trees $T_n$ with $n$ nodes where one unique node is designated as the root. Assume that there are $k$ nodes on the left subtree and $n - 1 - k$ nodes on the right subtree (so that the total number of nodes is $k + 1 + (n - 1 - k) = n$). It can be shown that it satisfies the recurrence

$$T_n = \sum_{k=0}^{n-1} T_k T_{n-1-k} \quad \text{for} \quad n \geq 1 \tag{2.15}$$

because $k$ can take any value from 0 (right skewed tree) to $n - 1$ (left skewed tree). The form of the recurrence suggests that it is of the form of some power (Table 2.2). But as the indexvar is varying from 0 to $n - 1$, it is not an exact power as shown below. Let $T(x) = \sum_{n=0}^{\infty} T_n x^n$. Multiply both sides of (2.15) by $x^n$ and sum over $n = 1$ to $\infty$ to get

$$\sum_{n=1}^{\infty} T_n x^n = \sum_{n=1}^{\infty} \sum_{k=0}^{n-1} T_k T_{n-1-k} x^n \quad \text{for} \quad n \geq 1, \tag{2.16}$$

where we have used $x$ instead of $t$ as dummy variable for convenience. As the LHS indexvar is varying from 1 onward, it is $T(x) - T_0$. Assume that $T_0 = 1$. Take one $x$ outside the summation on the RHS and write $x^{n-1} = x^k x^{n-1-k}$. Then the RHS is easily seen to be $x(T(x))^2$, so that we get $T(x) - 1 = x(T(x))^2$. Solve this as a quadratic equation in $T(x)$. This gives $T(x) = (1 \pm \sqrt{1 - 4x})/(2x)$. As $T(x) \to \infty$ as $x \to 0$, take negative sign to get $T(x) = (1 - \sqrt{1 - 4x})/(2x)$ as the required OGF for the number of binary trees with $n$ nodes. See Chap. 4 for another application of convolutions to matched parentheses.

**Table 2.2** Summary of convolutions and powers

| Type | Expression | Coefficient ($c_k$) |
|---|---|---|
| OGF Product (F(t)*G(t)) | $\sum_{i=0}^{\infty} a_i t^i * \sum_{j=0}^{\infty} b_j t^j$ | $\sum_{j=0}^{k} a_j b_{k-j}$ |
| OGF Power ($(F(t)^2)$) | $\left(\sum_{i=0}^{\infty} a_i t^i\right)^2$ | $\sum_{j=0}^{k} a_j a_{k-j}$ |
| OGF Product (of 3)(F(t)G(t)H(t)) | $\sum_{h=0}^{\infty} a_h t^h * \sum_{i=0}^{\infty} b_i t^i * \sum_{j=0}^{\infty} c_j t^j$ | $\sum_{h+i+j=k, k \geq 0} a_h b_i c_j$ |
| EGF Product ($H_1(t) * H_2(t)$) | $\sum_{i=0}^{\infty} a_i t^i / i! * \sum_{j=0}^{\infty} b_j t^j / j!$ | $\sum_{j=0}^{k} \binom{k}{j} a_j b_{k-j}$ |
| EGF Product ($H_1(t) * H_2(-t)$) | $\sum_{i=0}^{\infty} a_i t^i / i! * \sum_{j=0}^{\infty} b_j (-t)^j / j!$ | $\sum_{j=0}^{k} \binom{k}{j} (-1)^{k-j} a_j b_{k-j}$ |
| EGF Power ($[H_1(t)]^2$) | $\left(\sum_{i=0}^{\infty} a_i t^i / i!\right)^2$ | $\sum_{j=0}^{k} \binom{k}{j} a_j a_{k-j}$ |
| EGF Product (of 3) | $\sum_{h=0}^{\infty} a_h \frac{t^h}{h!} * \sum_{i=0}^{\infty} b_i \frac{t^i}{i!} * \sum_{j=0}^{\infty} c_j \frac{t^j}{j!}$ | $\sum_{h+i+j=k} a_h b_i c_j \frac{k!}{h!i!j!}$ |
| DGF Product | $\sum_{n=1}^{\infty} a_n/n^s * \sum_{n=1}^{\infty} b_n/n^s$ | $\sum_{d\mid n} a_d b_{n/d}$ |
| DGF Power | $\sum_{n=1}^{\infty} a_n/n^s * \sum_{n=1}^{\infty} a_n/n^s$ | $\sum_{j*k=n} a_j a_k$ |

### 9. OGF of Catalan numbers

Find the OGF of Catalan numbers that satisfy the recurrence $C_0 = 1, C_n = C_0 C_{n-1} + C_1 C_{n-2} + \cdots + C_{n-1} C_0$ for $n > 1$. Prove that Catalan numbers are always integers.

Let $F(t)$ denote the OGF of Catalan numbers. Write the convolution as a sum $C_n = \sum_{j=0}^{n-1} C_j C_{n-1-j}$. As per Table 2.2 entry row 2, this is the coefficient of a power. As done above, multiply by $t^n$ and sum over $0-\infty$ to get

$$\sum_{n=0}^{\infty} C_n t^n = \sum_{n=0}^{\infty} \sum_{j=0}^{n-1} C_j C_{n-1-j} t^n \quad \text{for} \quad n \geq 1. \tag{2.17}$$

Write $t^n = t^j t^{n-1-j} * t$. Then the RHS is easily found to be the coefficients of $t\, F(t)^2$. As $C_0 = 1$, we get $F(t) = t\, F(t)^2 + 1$. Solve this as a quadratic in $F(t)$ to get $F(t) = (1 \pm \sqrt{1-4t})/(2t)$. As $t \to 0$, $(1 + \sqrt{1-4t})/(2t) \to \infty$ (as it is 1/0 form), but the other root is 0/0 form, which by L'Hospitals rule tends to 1. Alternatively, the plus sign results in negative coefficients but as Catalan numbers are always positive integers, we have to take the minus sign. Hence, we take $F(t) = (1 - \sqrt{1-4t})/(2t)$, which is the same as in previous example. Expand $\sqrt{1+t}$ using

$$\sqrt{1+t} = (1+t)^{1/2} = \sum_{k=0}^{\infty} \binom{1/2}{k} t^k = 1 + \sum_{k=1}^{\infty} \binom{1/2}{k} t^k \tag{2.18}$$

because $\binom{1/2}{0} = 1$. Now use $\binom{1/2}{k} = 2^{1-2k}(-1)^{k-1}\binom{2k-2}{k-1}/k$ to get

$$\sqrt{1+t} = 1 + \sum_{k=1}^{\infty} 2^{1-2k}(-1)^{k-1}\binom{2k-2}{k-1}/k \, t^k. \tag{2.19}$$

Replace $t$ by $-4t$ and subtract from 1 to get

$$1 - \sqrt{1-4t} = -\sum_{k=1}^{\infty} 2^{1-2k}(-1)^{k-1}\binom{2k-2}{k-1}/k(-1)^k 2^{2k} t^k$$

$$= \sum_{k=1}^{\infty} 2\binom{2k-2}{k-1}/k \, t^k$$

as $(-1)^{k-1}(-1)^k = (-1)^{2k-1} = (-1)^{-1} = -1$ and $2^{1-2k}2^{2k} = 2$. Write $(2k-2)$ as $2(k-1)$ and replace $k-1$ by $k$ so that it varies from $0-\infty$ to get

$$\sum_{k=0}^{\infty} 2\binom{2k}{k}/(k+1) \, t^{k+1}. \tag{2.20}$$

Now divide by $2t$ to get $F(t) = \sum_{k=0}^{\infty}[1/(k+1)]\binom{2k}{k}t^k$, which is the required OGF. The quantity $[1/(n+1)]\binom{2n}{n}$ is called Catalan number. This can also be written as $\binom{2n}{n} - \binom{2n}{n+1}$. As $\binom{2n}{n+1}$ is always less than $\binom{2n}{n}$ (as can be seen from Pascal's triangle) this difference is always positive. Moreover, both of them are integers showing that Catalan numbers are always integers. The first few of them are 1, 1, 2, 5, 14, 42, 132, 429, .....

**10. Prove $\sum_{k=0}^{r} \binom{m}{k}\binom{n}{r-k} = \binom{m+n}{r}$**

If $m, n$, and $r$ are integers, prove the identity $\sum_{k=0}^{r} \binom{m}{k}\binom{n}{r-k} = \binom{m+n}{r}$.

As the LHS is the convolution of binomial coefficients, it suggests that it has OGF of the form $(1+t)^k$, for some integer $k$. We have shown in the last chapter that $(1+t)^m$ is the OGF of $\binom{m}{k}$. Hence, the OGF of LHS is $(1+t)^m(1+t)^n$ or equivalently $(1+t)^{m+n}$, which generates the RHS.

### 2.1.8  Differentiation and Integration

A univariate GF can be differentiated or integrated in the dummy variable (which is assumed to be continuous). Similarly, bivariate and higher GF can be differentiated or integrated in one or more dummy variables, as they are independent. See Table 2.3 for a summary of various operations.

**Table 2.3** Summary of generating functions operations

| Name | Expression | GF |
|------|-----------|-----|
| Add/subtract | $\sum_{k=0}^{\infty}(a_k \pm b_k)t^k$ | $F(t) \pm G(t)$ |
| Scalar multiply | $\sum_{k=0}^{\infty} ca_k$ | $c*F(t)$ |
| Convolution | $\sum_{k=0}^{\infty} c_k t^k$ | $F(t)*G(t)$ |
| Left shift (EGF) | $\sum_{k=0}^{\infty} a_{k+1} t^k/k!$ | $F'(t)$ |
| Left shift (OGF) | $\sum_{k=1}^{\infty} a_{k+1}\, t^k$ | $(F(t)-F(0))/t$ |
| Left m-shift(OGF) | $\sum_{k=0}^{\infty} a_{m+k}\, t^k$ | $(F(t)-a_0 - a_1 t \cdots - a_{m-1}t^{m-1})/t^m$ |
| Index multiply | $\sum_{k=0}^{\infty} ka_{k-1}t^k$ | $t\, F'(t)$ |
| Tail sum (OGF) | $\sum_{k=0}^{\infty} \sum_{j=0}^{k} a_j t^k$ | $F(t)/(1-t)$ |
| Difference (OGF) | $a_0 + \sum_{k=1}^{\infty}(a_k - a_{k-1})t^k$ | $(1-t)F(t)$ |

## Differentiation of OGF

Consider the OGF

$$F(t) = a_0 + a_1 t + a_2 t^2 + \cdots = \sum_{k=0}^{\infty} a_k t^k. \tag{2.21}$$

Differentiate w.r.t. $t$ to get

$$(\partial/\partial t)F(t) = a_1 + 2a_2 t + 3a_3 t^2 + \cdots = \sum_{k=1}^{\infty}(ka_k)t^{k-1}, \tag{2.22}$$

which is the OGF of $(a_1, 2a_2, 3a_3, \ldots)$ or $\{b_n = (n+1)a_{n+1}\}$ for $n \geq 0$. This can be stated as follows:

If $(a_0, a_1, a_2, \ldots) \Leftrightarrow F(t)$, then $(a_1, 2a_2, 3a_3, \ldots) \Leftrightarrow (\partial/\partial t)F(t) = F'(t)$. Literally, this means that differentiation of an OGF wrt the dummy variable multiplies each term by its index, and shifts the whole sequence left one place, so that $[t^{n-1}]F'(t) = n[t^n]F(t)$. Repeated differentiation gives the relationship $[t^{n-k}]F^{(k)}(t) = n_{(k)}[t^n]F(t)$ where $n_{(k)}$ is the falling Pochhammer notation. Next consider an EGF

$$H(t) = a_0 + a_1 t/1! + a_2 t^2/2! + \cdots = \sum_{k=0}^{\infty} a_k t^k/k!. \tag{2.23}$$

Differentiate w.r.t. $t$ to get

$$(\partial/\partial t)H(t) = a_1 + a_2 t/1! + a_3 t^2/2! + \cdots = \sum_{k=1}^{\infty} a_k t^{k-1}/(k-1)!, \tag{2.24}$$

because $k$ in the numerator cancels out with $k!$ in the denominator leaving a $(k-1)!$. This shows that differentiation of an EGF is equivalent to "shift-left" by one position.

**11. OGF of $t F'(t)$**

If $F(t)$ and $H(t)$ denote the OGF and EGF of a sequence $(a_0, a_1, a_2, \ldots)$ find the sequence whose OGF are, (i) $t F'(t)$ and (ii) $t H'(t)$.

The OGF of $F'(t)$ is found above as $(a_1, 2a_2, 3a_3, \cdots)$. Multiplying this by $t$ gives $t F'(t) = a_1 t + 2a_2 t^2 + 3a_3 t^3 + \cdots$. This can be written as $a_1 t + a_2 (\sqrt{2} t)^2 + a_3 (3^{1/3} t)^3 + \cdots$, in which the $k$th term is $a_k (k^{1/k} t)^k$. Next consider the EGF. As shown above, $H'(t)$ is the EGF of left-shifted sequence $(a_1, a_2, a_3, \ldots)$. Multiply both sides by $t$ to get

$$t(\partial/\partial t) H(t) = a_1 t + a_2 t^2 / 1! + a_3 t^3 / 2! + \cdots = \sum_{k=1}^{\infty} a_k / (k-1)! t^k, \qquad (2.25)$$

where we have associated $1/(k-1)!$ with the series term $a_k$. This is the OGF of $\{a_k/(k-1)!\}$.

### Higher-Order Derivatives

Differentiation and integration can be applied any number of times. For example, consider the formal power series $f(t) = 1/(1-t) = \sum_{k=0}^{\infty} t^k$. Take log of both sides and differentiate to get the first derivative as $f'(t)/f(t) = 1/(1-t)$ so that $f'(t) = 1/(1-t)^2$. Successive differentiation gives $f^{(n)}(t) = n!/(1-t)^{n+1}$. A direct consequence of this result is the GF

$$\sum_{k=0}^{\infty} \binom{n+k-1}{k} t^k = 1/(1-t)^n. \qquad (2.26)$$

Higher-order derivatives are used to find factorial moments of discrete random variables in the next chapter.

**12. OGF of $a_n = n^2$ using differentiation and shift operations**

Use differentiation and shift operations to prove that $t(1+t)/(1-t)^3$ is the GF of the squares $a_n = n^2$.

Consider the OGF $F(t) = (1/(1-t)^2) = \sum_{k=0}^{\infty} (k+1) t^k$. Multiply by $t$ to get $t/(1-t)^2 = \sum_{k=0}^{\infty} (k+1) t^{k+1}$. Differentiate the LHS to get $[(1-t)^2 + 2t(1-t)]/(1-t)^4$. The '$2t$' term cancels out resulting in $(1-t^2)/(1-t)^4$. Write $(1-t^2) = (1+t)(1-t)$ and cancel one (1-t) to get $(1+t)/(1-t)^3$. The RHS derivative is $\sum_{k=0}^{\infty} (k+1)^2 t^k$. Multiply LHS and RHS by $t$ to get the LHS as $t(1+t)/(1-t)^3$. Now RHS is the OGF of $n^2$.

## 13. $k$th derivative of OGF

Find the GF of the sequence $a_n = \binom{n+m}{m}$ for $m$ fixed.

Consider $F(x) = x^{n+m}/m!$. The first derivative is $F'(x) = (n+m)\,x^{n+m-1}/m!$. Repeat the differentiation $m$ times to get $F^{(k)}(x) = (n+m)(n+m-1)(n+m-2)\ldots nx^n/m!$. This can be written as $\binom{n+m}{m}x^n$. Thus, the GF is $(1/m!)(\partial/\partial x)^m(1/(1-x)) = 1/(1-x)^{m+1}$.

## 14. $k$th derivative of EGF

Prove that $D^k H(t) = \sum_{n=0}^{\infty} a_{n+k} t^n/n!$.

Consider $H(t) = a_0 + a_1 t/1! + a_2 t^2/2! + a_3 t^3/3! + \cdots + a_k t^k/k! + \cdots$. Take the derivative w.r.t. $t$ of both sides. As $a_0$ is a constant, its derivative is zero. Use derivative of $t^n = n * t^{n-1}$ for each term to get $H'(t) = a_1 + a_2 t/1! + a_3 t^2/2! + \cdots + a_k t^{k-1}/(k-1)! + \cdots$. Differentiate again (this time $a_1$ being a constant vanishes) to get $H''(t) = a_2 + a_3 t/1! + a_4 t^2/2! + \cdots + a_k t^{k-2}/(k-2)! + \cdots$. Repeat this process $k$ times. All terms whose coefficients are below $a_k$ will vanish. What remains is $H^{(k)}(t) = a_k + a_{k+1} t/1! + a_{k+2} t^2/2! + \cdots$. This can be expressed as $H^{(k)}(t) = \sum_{n=k}^{\infty} a_n t^{n-k}/(n-k)!$. Using the change of indexvar, this can be written as $H^{(k)}(t) = \sum_{n=0}^{\infty} a_{n+k} t^n/n!$. Now if we put $t = 0$, all higher-order terms vanish except the constant $a_k$.

## Integration

Integrating $F(t) = a_0 + a_1 t + a_2 t^2 + \cdots = \sum_{k=0}^{\infty} a_k t^k$ gives

$$\int_0^x F(t)dt = a_0 x + a_1 x^2/2 + a_2 x^3/3 + \cdots = \sum_{k \geq 1} (a_{k-1}/k)\, x^k. \qquad (2.27)$$

Divide both sides by $x$ to get

$$(1/x) \int_0^x F(t)dt = (a_0/1) + (a_1/2)x + (a_2/3)x^2 + \cdots \qquad (2.28)$$

which is the OGF of the sequence $\{a_k/(k+1)\}_{k \geq 0}$, so that $[t^n] \int_t F(t)dt = (1/n)[t^{n-1}]F(t)$.

Next consider the EGF

$$H(t) = a_0 + a_1 t/1! + a_2 t^2/2! + a_3 t^3/3! + \cdots = \sum_{k=0}^{\infty} a_k t^k/k!. \qquad (2.29)$$

Integrate term by term to get

$$\int_0^x H(t)dt = a_0 x + a_1 x^2/(1! * 2) + a_2 x^3/(2! * 3) + \cdots = \sum_{k=0}^{\infty} a_k x^{k+1}/(k+1)!. \qquad (2.30)$$

Integration of EGF is equivalent to shift-right by one position.

### 15. $\sum_{k=0}^{n} \binom{n}{k}/(k+1)$

Evaluate the sum $\sum_{k=0}^{n} \binom{n}{k}/(k+1)$.

Consider the OGF $\sum_{k=0}^{n} \binom{n}{k}/(k+1)x^k$. With the help of above result, this can be written as an integral $(1/x) \int_0^x (1+t)^n dt$. As the integral of $(1+t)^n$ is $(1+t)^{n+1}/(n+1)$, we get $[(1+x)^{n+1} - 1]/[(n+1)x]$ as the result.

### 16. OGF of 1, 1/3, 1/5, 1/7, …

Find the OGF of the sequence 1, 1/3, 1/5, 1/7, ….

Consider the sequence $F(x) = 1 + x^2 + x^4 + x^6 + \cdots$, which has OGF $(1-x^2)^{-1}$. Integrate the RHS term-by-term to get $x + x^3/3 + x^5/5 + x^7/7 + \cdots$, which has the desired coefficients. As the LHS is $\int (1-x^2)^{-1} dx$, either put $x = \sin(\theta)$, $dx = \cos(\theta)d\theta$, or use partial fractions to break this into two terms $A/(1-x) + B/(1+x)$. This gives $A + B = 1$ and $A - B = 0$, from which $A = B = 1/2$. Thus $(1-x^2)^{-1} = (1/2)[1/(1-x) + 1/(1+x)]$. Now integrate term-by-term on the RHS to get $\frac{1}{2}[-\log(1-x) + \log(1+x)] = \frac{1}{2}\log((1+x)/(1-x))$, using $\log(a) - \log(b) = \log(a/b)$. This is the OGF of the original sequence, which is of type LogGF.

---

## 2.2   Multiplicative Inverse Sequences

Two GFs $F(t)$ and $G(t)$ are invertible if $F(t)G(t) = 1$. The $G(t)$ is called the "multiplicative inverse" of $F(t)$. Consider the sequence $(a_0, a_1, \ldots)$. Let $G(x)$ be the multiplicative-inverse with coefficients $(b_0, b_1, \ldots)$, so that $F(t)G(t) = 1$. This gives $c_0 = a_0 b_0$, from which $b_0 = 1/a_0$. In general, $b_n = -(\sum_{k=1}^{n} a_k b_{n-k})/a_0$. If $a_0 \neq 0$, it is invertible and coefficients can be calculated recursively. This shows that a necessary condition for an OGF to be "invertible" is that the constant term is nonzero.

### 17. Inverse of $(1, 1, 1, \ldots)$

Find the inverse of the sequence $(1, 1, 1, \ldots)$.

Let $G(t)$ be the inverse with coefficients $(b_0, b_1, \ldots)$. Then $a_0 b_0 = 1$ implies that $b_0 = 1$. Similarly, $a_0 b_1 + a_1 b_0 = 0$ implies that $1 * b_1 + 1 * 1 = 0$ or $b_1 = -1$. All other coefficients are zeros. For example, in $b_2 + b_1 + b_0 = 0$ put $b_0 = 1$ and $b_1 = -1$, so that $b_2 = 0$. Similarly, all higher coefficients are zeros. Hence, the inverse is $(1, -1, 0, 0, \ldots)$.

## 2.3 Composition of Generating Functions

If $F(x)$ and $G(x)$ are two GFs, the composition is defined as $F(G(x))$. Note that $F(G(x))$ is not the same as $G(F(x))$. Consider $F(x) = 1/(1-x)$ and $G(x) = x(1+x)$. Then $F(G(x))$ is $1/(1-x(1+x))$. The general result of composition requires higher powers of GF. Nevertheless, if we make the assumption that $F(G(x)) = G(F(x)) = x$, the coefficient of compositions of OGFs can be found easily. Suppose $F(x) = \sum_{n=0}^{\infty} a_n x^n$ and $G(x) = \sum_{m=0}^{\infty} b_m x^m$. To find $F(G(x))$, we have to substitute $G(x)$ in place of $x$ in $F(x)$. This gives $H(x) = F(G(x)) = \sum_{n=0}^{\infty} a_n (\sum_{m=0}^{\infty} b_m x^m)^n$. If $H(x) = \sum_{n=0}^{\infty} c_n x^n$, the constant $c_0$ is obtained by equating to the constant term on the RHS. This gives $c_0 = a_0 + a_1 b_0 + a_2 b_0^2 + \cdots + a_k b_0^k + \cdots$. As $F(G(x)) = x, c_0 = 0$. The RHS can be zero only if $a_0 = -(a_1 b_0 + a_2 b_0^2 + \cdots + a_k b_0^k + \cdots)$ or $a_0 = b_0 = 0$. If we assume $a_0 = b_0 = 0$, higher order coefficients can easily be found as $c_1 = a_1 b_1, c_2 = a_1 b_2 + a_2 b_1^2$, etc.

If both of them are OGF, then both compositions are also OGF type. In all other cases, it is of EGF type. For instance, if $F(x)$ is an OGF and $H(x)$ is an EGF, both $F(H(x))$ and $H(F(x))$ are of type EGF.

## 2.4 Summary

This chapter introduced various operations on GFs. This includes arithmetic operations, linear combinations, left and right shifts, convolutions, powers, change of dummy variables, differentiation and integration. These find applications in stack filters used in DSP (Wendt et.al. 1986; Kuosmanen et.al. 1994),

## References

Chouraqui, F. (2019). About an extension of the Davenport-Rado result to the Herzog_Schönheim conjecture for free groups, preprint. arXiv:1901.09898v1.

Graham, R., Knuth, D. E., & Patashnik, O. (1994). *Concrete mathematics* (2nd ed.). MA: Addison Wesley.

Kuosmanen, P., Astola, J., & Agaian, S. (1994). On rank selection probabilities. *IEEE Transactions on Signal Processing, 42*(11), 3255–3258.

Wendt, P., Coyle, E., & Gallagher, N. (1986). Stack filters. *IEEE Transactions on Acoustics, Speech, and Signal Processing, 34,* 898–911.

Wilf, H. (1994). *Generating functionology*. Academic.

# Generating Functions in Statistics

**3**

This chapter discusses most popular generating functions (GFs) in statistics like probability GFs; CDF GFs; moment and cumulant GFs, characteristic functions. Some new GFs like mean deviation GF and survival function GF are discussed. A discussion of factorial moment GF and its use in finding the factorial moments of binomial and Poisson distribution is also included. Conditional moment GF and GF for truncated distributions are briefly introduced.

## 3.1 Uses of Generating Functions in Statistics

GFs are used in various branches of statistics like distribution theory, stochastic processes, etc. A one-to-one correspondence is established between the power series expansion of a GF in one or more auxiliary (dummy) variables, and the coefficients of a known sequence. These coefficients differ in various GFs as shown below. GFs used in discrete distribution theory are defined on the sample space of a random variable or probability distribution. They can be finite or infinite, depending on the corresponding distribution. Multiple GFs like PGF, MGF, etc. can be defined on the same random variable, as these generate different quantities as shown in the following paragraphs. Some of these GFs can be obtained from others using various transformations discussed in Chap. 2. The GF technique is especially useful in some distributions in investigating inter-relationships and asymptotics, sums and limits of random variables, and in characterizations of distributions. As an example, consider the Poisson distribution with parameter $\lambda$. If the average number of occurrences of a rare event in a certain time interval $(t, t + dt)$ is $\lambda$, the Poisson probabilities gives us the chance of observing exactly $k$ events in the same time interval, if various events occur independently of each other. The PGF of Poisson distribution is infinitely differentiable, and it is easy to

© The Author(s), under exclusive license to Springer Nature Switzerland AG 2023     45
R. Chattamvelli and R. Shanmugam, *Generating Functions in Engineering and the Applied Sciences*, Synthesis Lectures on Engineering, Science, and Technology,
https://doi.org/10.1007/978-3-031-21143-0_3

find factorial moments because the $k$th derivative of PGF is $\lambda^k$ times the PGF. The GFs can be used to prove the additivity property of Poisson, negative binomial, and many other distributions (Chattamvelli and Shanmugam 2020). Although the previous chapter used $x$ as the dummy variable in a GF, we will use $t$ as the dummy variable in this chapter because $x$ is regarded as a random variable, and used as a subscript.

### 3.1.1  Types of Generating Functions

There are four popular GFs used in statistics—namely (i) probability generating function (PGF), denoted by $P_x(t)$; (ii) moment generating function (MGF), denoted by $M_x(t)$; (iii) cumulant generating function (CGF), denoted by $K_x(t)$; and (iv) characteristic function (ChF), denoted by $\phi_x(t)$. In addition, there are still others like factorial moment GF (FMGF), inverse moment GF (IMGF), inverse factorial moment GF (IFMGF), absolute moment GF, as well as GFs for odd moments and even moments separately. When the variance of a discrete random variable is linear in one of the (integer) parameters, it is possible to define a variance generating function (VGF). As an example, the VGF of binomial distribution BINO(n,p) is given by $V_x(t; n, p) = pq/(1 - t)^2$ (Chattamvelli and Shanmugam 2020). These are called "canonical functions" in some fields.

The PGF generates the probabilities of a random variable, and is of type OGF. The rising and falling FMGFs are also of type OGF. The MGF generates moments, and is of type EGF. It has further subdivisions as ordinary MGF, central MGF, FMGF, IMGF, and IFMGF. The CGF and ChF are also related to MGF. Some of these (MGF, ChF, etc.) can also be defined for an *arbitrary origin*. The CGF is defined in terms of the MGF as $K_x(t) = \ln(M_x(t))$,

**Table 3.1**  Summary table of generating functions

| Abbreviation | Symbol | Definition E = expectation oper. | Generates what | How obtained (t = dummy-variable) |
|---|---|---|---|---|
| PGF | $P_x(t)$ | $E(t^x)$ | Probabilities | $p_k = \frac{\partial^k}{\partial t^k} P_x(t)\vert_{t=0}/k!$ |
| CDFGF | $F_x(t)$ | $E(t^x/(1\text{-}t))$ | Cumulative prob. | $F_k = \frac{1}{k!}\frac{\partial^k}{\partial t^k} P_x(t)/(1 - t)\vert_{t=0}$ |
| SFGF | $S_x(t)$ | $E((1 - t^x)/(1\text{-}t))$ | Cumulative Tail prob. | $S_k = \frac{1}{k!}\frac{\partial^k}{\partial t^k}(1 - P_x(t))/(1 - t)\vert_{t=0}$ |
| MGF | $M_x(t)$ | $E(e^{tx})$ | Moments | $\mu_k' = \frac{\partial^k}{\partial t^k} M_x(t)\vert_{t=0}$ |
| CMGF | $M_z(t)$ | $E(e^{t(x-\mu)})$ | Central moments | $\mu_k = \frac{\partial^k}{\partial t^k} M_z(t)\vert_{t=0}$ |
| ChF | $\phi_x(t)$ | $E(e^{itx})$ | Moments | $i^k \mu_k' = \frac{\partial^k}{\partial t^k} \phi_x(t)\vert_{t=0}$ |
| CGF | $K_x(t)$ | $\log(E(e^{tx}))$ | Cumulants | $\mu_k = \frac{\partial^k}{\partial t^k} K_x(t)\vert_{t=0}$ |
| FMGF | $\Gamma_x(t)$ | $E((1 + t)^x)$ | Factorial moments | $\mu_{(k)} = \frac{\partial^k}{\partial t^k} \Gamma_x(t)\vert_{t=0}$ |
| MDGF | $D_x(t)$ | $2E(t^x/(1\text{-}t)^2)$ | Mean deviation | see Sect. 3.5 |

Probability generating function (PGF) is of type OGF. MGF and ChF are of type EGF. MGF need not always exist, but characteristic function always exists. Falling factorial moment is denoted as $\mu_{(k)} = E(x(x - 1)(x - 2) \cdots (x - k + 1))$

which when expanded as a polynomial in t gives the cumulants. Note that the logarithm is to the base $e(\ln)$. As every distribution does not possess a MGF, the concept is extended to the complex domain by defining the ChF as $\phi_x(t) = E(e^{itx})$. If all of them exists for a distribution, then

$$P_x(e^t) = M_x(t) = e^{K_x(t)} = \phi_x(it). \tag{3.1}$$

This can also be written in the alternate forms $P_x(e^{it}) = M_x(it) = e^{K_x(it)} = \phi_x(-t)$ or as $P_x(t) = M_x(\ln(t)) = e^{K_x(\ln(t))} = \phi_x(i \ln(t))$ (Johnson et.al. (2005)) (Table 3.1).

## 3.2   Probability Generating Functions (PGF)

The PGF is extensively used in statistics, econometrics, and various engineering fields. It is a compact mathematical expression in one or more dummy variables, along with the unknown parameters, if any, of a distribution.

**Definition 3.1**   A GF in which the coefficients are the probabilities associated with a random variable is called the PGF.

As mentioned in Chap. 1, the PGF is not a function that generates probabilities when particular values are plugged in for the dummy variable $t$. But when expanded as a power series in the auxiliary variable, the coefficient of $t^k$ is the probability of the random variable $X$ taking the value $k$. This is one way the PGF of a random variable can be used to generate probabilities. It is defined as

$$P_x(t) = E(t^x) = \sum_x t^x p(x) = p(0) + p(1)\,t + p(2)\,t^2 + \cdots + p(k)\,t^k + \cdots, \tag{3.2}$$

where the summation is over the range of $X$. This means that the PGF of a discrete random variable is the weighted sum of all probabilities of outcomes $k$ with weight $t^k$ for each value of $k$ in its range. Obviously the RHS of (3.2) is a finite series for distributions with finite range. It usually results in a compact closed form expression for discrete distributions. It converges for $|t| < 1$, and appropriate derivatives exist. Differentiating both sides of (3.2) $k$ times w.r.t. $t$ gives $(\partial^k/\partial t^k)P_x(t) = k!p(k)+$ terms involving $t$. If we put $t = 0$, all higher order terms that have "$t$" or higher powers vanish, giving $k!p(k)$. From this $p(k)$ is obtained as $(\partial^k/\partial t^k)P_x(t)|_{t=0}/k!$ (see Table 3.1). If the $P_x(t)$ involves powers or exponents, we take the log (w.r.t. $e$) of both sides and differentiate $k$ times, and then use the following result on $P_x(t = 1)$ to simplify the differentiation. The PGF is immensely useful in deriving key properties of a random variable easily. As shown below, it is related to CDFGF and MDGF.

**1. PGF special values $P_x(t = 0)$ and $P_x(t = 1)$**

Find $P_x(t = 0)$ and $P_x(t = 1)$ from the PGF of a discrete distribution.

As $\sum_k p(k)$ being the sum of the probabilities is one, it follows trivially by putting $t = 1$ in (3.2) that $P_x(t = 1) = 1$. Put $t = 0$ in (3.2) to get $P_x(t = 0) = p(0)$, the first probability. Similarly, put $t = -1$ to get the RHS as $[p(0) + p(2) + \cdots +] - [p(1) + p[3] + p[5] + \cdots ]$.

**2. Finding the mean from PGF**

Find the PGF of a random variable given below, and obtain its mean.

| $x$    | 1   | 2   | 3   |
|--------|-----|-----|-----|
| $p(x)$ | 1/2 | 1/3 | 1/6 |

Plug in the values directly in

$$P_x(t) = E(t^x) = \sum_{x=1}^{3} t^x p(x) \tag{3.3}$$

to get $P_x(t) = t/2 + t^2/3 + t^3/6$. Differentiate w.r.t. $t$ and put $t = 1$ to get the mean as $1/2 + 2t/3 + 3t^2/6|_{t=1} = 1/2 + 2/3 + 3/6 = 5/3$.

**3. PGF of uniform distribution**

Find PGF of a discrete uniform distribution, and obtain (i) the difference between the sum of even and odd probabilities, and (ii) the mean.

Take $f(x) = 1/k$ for $x = 1, 2, \ldots k$ for simplicity. Then $P_x(t) = 1/k[t^1 + t^2 + \cdots + t^k]$. Take $t$ outside, and apply formula for geometric progression to get $P_x(t) = (t/k)(t^k - 1)/(t - 1)$. Put $t = -1$ to get the answer to (i). Take log and differentiate. Then put $t = 0$, to get the mean as $(k + 1)/2$.

### 4. PGF of Poisson distribution

Find the PGF of a Poisson distribution, and obtain the difference between the sum of even and odd probabilities.

The PGF of a Poisson distribution is

$$P_x(t) = E(t^x) = \sum_{x=0}^{\infty} t^x e^{-\lambda} \lambda^x / x! = e^{-\lambda} \sum_{x=0}^{\infty} (\lambda t)^x / x! = e^{-\lambda} e^{t\lambda} = e^{-\lambda[1-t]}. \quad (3.4)$$

Put $t = -1$ in (3.4) and use the above result to get the desired sum as $\exp(-\lambda[1 - (-1)]) = \exp(-2\lambda)$.

### 5. PGF of geometric distribution

Find the PGF of a geometric distribution with PMF $f(x; p) = q^x p, x = 0, 1, 2 \ldots$, and obtain the difference between the sum of even and odd probabilities.

As $x = 0, 1, 2, \ldots \infty$ values, we get the PGF as

$$P_x(t) = E(t^x) = \sum_{x=0}^{\infty} t^x q^x p = p \sum_{x=0}^{\infty} (qt)^x = p/(1 - qt). \quad (3.5)$$

Now $P[X \text{ is even}] = q^0 p + q^2 p + \cdots = p[1 + q^2 + q^4 + \cdots] = p/(1 - q^2) = 1/(1 + q)$, and $P[X \text{ is odd}] = q^1 p + q^3 p + \cdots = qp[1 + q^2 + q^4 + \cdots] = qp/(1 - q^2) = q/(1 + q)$. Using the above result, the difference between these must equal the value of $P_x(t = -1)$. Put $t = -1$ in (3.5) to get $p/(1 - q(-1)) = p/(1 + q)$, which is the same as $1/(1 + q) - q/(1 + q) = p/(1 + q)$. There is another version of geometric distribution with range $1-\infty$ with PMF $f(x; p) = q^{x-1} p$ (Chattamvelli and Shanmugam 2020). In this case the PGF is $pt/(1 - qt)$.

Closed-form expressions for $P_x(t)$ are available for most of the common discrete distributions. They are seldom used for continuous distributions because $\int t^x f(x) dx$ may not be convergent.

### 6. PGF of BINO$(n, p)$ distribution

Find the PGF of BINO$(n, p)$ distribution with PMF $f(x; n, p) = \binom{n}{x} p^x q^{n-x}$, and obtain the mean.

By definition

$$P_x(t) = E(t^x) = \sum_{x=0}^{n} \binom{n}{x} p^x q^{n-x} t^x = \sum_{x=0}^{n} \binom{n}{x} (pt)^x q^{n-x} = (q + pt)^n. \quad (3.6)$$

The coefficient of $t^x$ gives the probability that the random variable takes the value $x$. To find the mean, we take log of both sides. Then $\log(P_x(t)) = n * \log(q + pt)$. Differentiate both sides w.r.t. $t$ to get $P'_x(t)/P_x(t) = n * p/(q + pt)$. Now put $t = 1$ and use $P_x(t = 1) = 1$ to get the RHS as $n * p/(q + p) = np$ as $q + p = 1$.

### 7. PGF of NBINO($k$, $p$) distribution

Find the PGF of negative binomial distribution NBINO($k$, $p$), and obtain the mean.
   By definition

$$P_x(t) = E(t^x) = \sum_{x=0}^{\infty} \binom{x + k - 1}{x} p^k q^x t^x = \sum_{x=0}^{\infty} \binom{x + k - 1}{x} p^k (qt)^x. \qquad (3.7)$$

As $p^k$ is a constant, this is easily seen to be $[p/(1 - qt)]^k$. Take log and differentiate w.r.t. $t$ and put $t = 1$ to get $E(X) = kq/p$. By setting $p = 1/(1 + r)$ and $q = 1 - p = r/(1 + r)$ we get another form of negative binomial distribution NBINO($k$, $1/(1 + r)$) as $\binom{x+k-1}{x}[1/(1 + r)]^k[r/(1 + r)]^x$ with PGF $1/[1 + r(1 - t)]^k$. The PGF of NBINO($k$, $p$) can directly be derived by putting $x = tq = t(1 - p)$ in the expansion $1/(1 - x)^k$ and multiplying both sides by $p^k$.

### 8. PGF of logarithmic distribution

Find the PGF of logarithmic($q$) distribution, and obtain the mean.
   By definition

$$P_x(t) = - \sum_{x=1}^{\infty} q^x t^x /(x \log(p)) = (-1/\log(p)) \sum_{x=1}^{\infty} (qt)^x /x = \log(1 - qt)/\log(1 - q),$$

$$(3.8)$$

where $q = 1 - p$ and $t < 1/q$ for convergence. This distribution is a special limiting case of negative binomial distribution when the zero class is dropped, and the parameter $k$ of negative binomial distribution $\rightarrow 0$.

$$\left[ kp^k/(1 - p^k) \right] \left[ 0, q, (k + 1)q^2/2! + (k + 1)(k + 2)q^3/3! + \cdots \right]. \qquad (3.9)$$

When $k \rightarrow 0$, a factorial term will remain in the numerator which cancels out with the denominator, leaving a $k$ in the denominator. For example, $(k + 1)(k + 2)q^3/3!$ becomes $(0 + 1)(0 + 2)q^3/3! = 1.2q^3/3! = q^3/3$. Divide the numerator and denominator of the term outside the bracket by $p^k$ to get $k/p^{-k} - 1$. Use L'Hospitals rule once to get this limit as $-1/(\ln(p))$. Thus, the negative binomial distribution tends to the logarithmic law in the limit. Differentiate the PGF w.r.t. $t$ to get $\partial/\partial t P_x(t) = (-1/\log(p)) - q/(1 - qt)$. Put $t = 1$ to get the mean as $q/[\log(p)(1 - q)] = q/[p \log(p)]$.

### 9. PGF of hypergeometric distribution

Find the PGF of a hypergeometric distribution $\binom{m}{k}\binom{N-m}{n-k}/\binom{N}{n}$.

The PGF is easily obtained as

$$P_x(t) = \sum_{x=0}^{m} \binom{m}{x}\binom{N-m}{n-x}/\binom{N}{n} t^x \tag{3.10}$$

because the probabilities outside the range $(x > m)$ are all zeros. The hypergeometric series is defined in terms of rising Pochhammer EGF of $a^{(k)}b^{(k)}/c^{(k)}$ as $F(x; a, b, c) = \sum_k \frac{a^{(k)}b^{(k)}}{c^{(k)}} \frac{x^k}{k!}$ where $a^{(k)}$ denotes the rising Pochhammer number. A property of the hypergeometric series is that the ratio of two consecutive terms is a polynomial. More precisely, the ratio of $(k+1)$th to $k$th term is a polynomial in $k$. If such a polynomial is given, we could easily ascertain the parameters of the hypergeometric function from it. Consider the ratio of two successive terms of hypergeometric distribution

$$f(k+1)/f(k) = (k-m)(k-n)/[(k+N-m-n+1)(k+1)] \tag{3.11}$$

which is always written with "$k$" as the first variable. Now drop each $k$, and collect the remaining values to find out the corresponding hypergeometric function parameters as $F(-m, -n; N - m - n + 1; x)$. From the PMF we have that $f(0) = \binom{N-m}{n}/\binom{N}{n}$. This should be used as multiplier of the hypergeometric function to get $f(x) = \binom{N-m}{n}/\binom{N}{n}F(-m, -n; N - m - n + 1; x)$. From this we get the PGF as $\binom{N-m}{n}/\binom{N}{n}{}_2F_1(-m, -n; N - m - n + 1; t)$.

### 10. PGF of power-series distribution

Find the PGF of power-series distribution, and obtain the mean and variance.

The PMF of power-series distribution is $P(X = k) = a_k\theta^k/F(\theta)$ where $k$ takes any integer values. By definition

$$P_x(t) = (1/F(\theta)) \sum_{x=1}^{\infty} a_x\theta^x t^x = (1/F(\theta)) \sum_{x=1}^{\infty} a_x(\theta t)^x = F(\theta t)/F(\theta). \tag{3.12}$$

Differentiate wrt $t$ to get $P_x'(t) = \theta F'(\theta t)/F(\theta)$, from which the mean $\mu = P_x'(t)|_{t=1} = \theta F'(\theta)/F(\theta)$. The variance is found using $V(X) = E(X(X-1)) + E(X) - [E(X)]^2$. Consider $F''(t) = \theta^2 F''(\theta t)/F(\theta)$. From this the variance is obtained as $F''(1) + F'(1) - [F'(1)]^2 = \theta^2 F''(\theta)/F(\theta) + \theta F'(\theta)/F(\theta) - [\theta F'(\theta)/F(\theta)]^2$.

## 3.2.1   Properties of PGF

1. $P^{(r)}(0)/r! = \partial^r/\partial t^r \, P_x(t)|_{t=0} = P[X = r]$.

   $P_x(t)$ is infinitely differentiable in $t$ for $|t| < 1$. Differentiate $P_x(t) = E(t^x)$ w.r.t $t$ $r$ times to get $\frac{\partial^r}{\partial t^r} P_x(t) =$

$$E\left[x(x-1)\ldots(x-r+1)t^{x-r}\right] = \sum_{x \geq r}\left[x(x-1)\ldots(x-r+1)t^{x-r}\right] f(x). \quad (3.13)$$

   The first term in this sum is $[r(r-1)\ldots(r-r+1)t^{r-r}]f(x=r) = [r!t^0]f(x) = r!f(x=r)$. By putting $t = 0$, every term except the first vanishes, and the RHS becomes $r!f(x=r)$. Thus, $\frac{\partial^r}{\partial t^r} P_x(t=0) = r!f(x=r)$.

2. $P^{(r)}(1)/r! = E[X_{(r)}]$, the $r$th falling factorial moment (Sect. 3.10). By putting $t = 1$ in (3.13), the RHS becomes $E[x(x-1)\ldots(x-r+1)]$, which is the $r$th falling factorial moment. This is sometimes called the FMGF (see Sect. 3.10).

3. $\mu = E(X) = P'(1)$, and $\mu'_2 = E(X^2) = P''(1) + P'(1)$.

   The first result follows directly from the above by putting $r = 1$. As $X^2 = X(X-1) + X$, the second result also follows from it.

4. $V(X) = P''(1) + P'(1)[1 - P'(1)]$.

   This result follows from the fact that $V(X) = E[X^2] - E[X]^2 = E[X(X-1)] + E[X] - E[X]^2$. Now use the above two results.

5. $\int_t P_x(t)dt|_{t=1} = E(\frac{1}{X+1})$.

   Consider $\int_t P_x(t)dt = \int_t E(t^x)dt$. Take expectation operation outside the integral and integrate to get the result. This is the first inverse moment  (of $X + 1$), and holds for positive random variables.

6. $P_{cX}(t) = P_X(t^c)$.

   This follows by writing $t^{cX}$ as $(t^c)^X$.

7. $P_{X \pm c}(t) = t^{\pm c} * P_x(t)$.

   This follows by writing $t^{X \pm c}$ as $(t^{\pm c})t^X$.

8. $P_{cX \pm d}(t) = t^{\pm d} * P_X(t^c)$.

   This follows by combining two cases above.

9. $P_x(t) = M_X(\ln(t))$.

   From (3.1), we have $P_x(e^t) = M_x(t)$. Write $t' = e^t$ so that $t = \ln(t')$ to get the result.

10. $P_{(X \pm \mu)/\sigma}(t) = t^{\pm \mu/\sigma} P_X(t^{1/\sigma})$.

    This is called the change of origin and scale transformation of PGF. This follows by combining (6) and (7).

11. If $X$ and $Y$ are IID random variables, $P_{(X+Y)}(t) = P_x(t)P_y(t)$ and $P_{(X-Y)}(t) = P_x(t)P_Y(1/t)$.

PGF of the absolute difference of two non-negative random variables can be expressed as an integral $G_{|X-Y|}(t) = \frac{1}{2\pi} \int_0^{2\pi} \frac{1-t^2}{1+t^2-2t\cos(\theta)} G_{x,y}(\exp(i\theta), \exp(-i\theta))d\theta$  (Puri 1966).

## 3.3    Generating Functions for CDF

As the PGF of a random variable generates probabilities, it can be used to generate the sum of left tail probabilities (CDF).

**Definition 3.2**  A GF in which the coefficients are the cumulative probabilities associated with a random variable is called the CDF generating function (CDFGF).

We have seen in Theorem 1 (Sect. 1.3.3) that if $F(x)$ is the OGF of the sequence $(a_0, a_1, \ldots, a_n, \ldots)$, finite or infinite, then $F(x)/(1-x)$ is the OGF of the sequence $(a_0, a_0 + a_1, a_0 + a_1 + a_2, \ldots)$. By replacing $a_i$'s by probabilities, we obtain a GF that generates the sum of probabilities as

$$G(t) = \sum_{k=0}^{\infty} \left( \sum_{j=0}^{k} p_j \right) t^k = p_0 + (p_0 + p_1)t + (p_0 + p_1 + p_2)t^2 + \cdots. \qquad (3.14)$$

This works only for discrete distributions.

### 3.3.1   Properties of CDFGF

1. Particular cases: $G_x(0) = p_0$, $G_x'(0) = p_0 + p_1$, and $G_x^{(n)}(0) = n! \sum_{k=0}^{n} p_k$ where $G_x^{(n)}(0)$ is the $n$th derivative wrt t.
2. $G_{x\pm c}(t) = t^{\pm c} G_x(t)$
3. If X and Y are independent, $G_{x+y}(t) = G_x(t)P_y(t) = P_x(t)G_y(t)$.
4. If X and Y are independent, $G_{x-y}(t) = G_x(t)P_y(1/t)$.

**11. CDFGF of geometric distribution**

Obtain the CDFGF of a geometric distribution with PMF $f(x; p) = q^x p, x = 0, 1, 2 \ldots$.
     The PGF of a geometric distribution is derived in Page 49 as $p/(1 - qt)$ from which the CDFGF is obtained as $G(t) = p(1-t)^{-1}/(1 - qt)$. Expand both $(1-t)^{-1}$ and $(1-qt)^{-1}$ as infinite series' and combine like powers to get

$$G(t) = p\left[1 + t(1+q) + t^2\left(1 + q + q^2\right) + t^3\left(1 + q + q^2 + q^3\right) + \cdots\right]. \qquad (3.15)$$

Write $(1 + q + q^2 + q^3 + \cdots + q^k)$ in (3.15) as $(1 - q^{k+1})/(1 - q)$, and cancel $(1 - q) = p$ in the numerator to get the CDFGF of geometric distribution as

$$D_x(t) = \left[ 1 + t\left(1 - q^2\right) + t^2\left(1 - q^3\right) + t^3\left(1 - q^4\right) + \cdots \right]. \qquad (3.16)$$

The CDF of geometric distribution is related to the incomplete beta function as $I_p(1, k)$ for $k = 1, 2, 3, \ldots$. This is useful to mitigate underflow problems in computer memory when $k$ is very large (CDF is sought in extreme right tail). As $I_p(1, k) = 1 - I_q(k, 1)$, where q=1-p, this can also be used to speed-up the computation. Geometric CDF can be used to model exceedences of a specified magnitude in hydrology. For example, maximum dam discharges from reservoirs can be modeled where p = P(X$\geq$ $x_T$) is the occurrence of the event. The interval between two such exceedences is called return period in hydrology and geology.

### 12. CDFGF of Poisson distribution

Obtain the CDFGF of a Poisson distribution with $f(x; \lambda) = e^{-\lambda}\lambda^x/x!$, x=0,1,$\cdots\infty$.

The PGF of Poisson distribution is found in Page 49 as $e^{-\lambda[1-t]}$. From this the CDFGF follows easily as $e^{-\lambda(1-t)}/(1 - t)$. Write the numerator as $e^{-\lambda}e^{\lambda t}$. Now expand $e^{\lambda t}$ and $(1 - t)^{-1}$ as infinite series and collect coefficients of like powers to get

$$D_x(t) = \left[ 1 + t\left(e^{-\lambda}\right) + t^2/2!\left(e^{-\lambda}[1 + \lambda]\right) + t^3/3!\left(e^{-\lambda}\left[1 + \lambda + \lambda^2/2!\right]\right) + \cdots \right].$$
$$(3.17)$$

### 13. CDFGF of negative binomial distribution

Obtain the CDFGF of a negative binomial distribution with PMF $f(x; k, p) = \binom{x+k-1}{x}p^k q^x$, $x = 0, 1, 2, \ldots$.

The PGF of negative binomial distribution is derived in Page 50 as $[p/(1 - qt)]^k$ from which the CDFGF is obtained as $G(t) = p^k(1 - t)^{-1}(1 - qt)^{-k}$. As the CDF of negative binomial distribution is $I_p(n, k + 1)$ for $k = 0, 1, 2, \ldots$ this can be used when $k$ is very large.

## 3.4    Generating Function for Survival Probabilities

As the PGF of a random variable generates probabilities, it can be used to generate the sum of right tail probabilities, called survival function (SF).

**Definition 3.3**  A GF in which the coefficients are the survival probabilities associated with a random variable is called the SF generating function (SFGF).

First consider a finite distribution with PMF $f(x)$ and range 0–$n$. Let $a_0, a_1, \ldots, a_n$ be the corresponding probabilities (coefficients). As $t^n P_x(1/t)$ reverses the probabilities, $t^n P_x(1/t)/(1-t)$ is the SFGF. The SF is defined as $P[X > k]$ (where the $x = k$ case is not included) when the CDF is defined as $P[X \le k]$. For distributions with infinite range, the SFGF is $(1 - P_x(t))/(1-t)$. Intuitively, $(1 - P_x(t))$ generates complementary (tail) probabilities and multiplying it by $1/(1-t)$ results in the sum of these probabilities as per Theorem 1 (Sect. 1.3.3).

## 14. SFGF of BINO($n, p$) distribution

Find the SFGF of BINO($n, p$) distribution with PMF $f(x; n, p) = \binom{n}{x} p^x q^{n-x}$, x=0, 1, $\cdots$, n.

By definition

$$P_x(t) = E(t^x) = \sum_{x=0}^{n} \binom{n}{x} p^x q^{n-x} t^x = \sum_{x=0}^{n} \binom{n}{x} (pt)^x q^{n-x} = (q + pt)^n. \qquad (3.18)$$

Using the above result, the SFGF follows as $t^n/(1-t)$ $(q + p/t)^n$. Taking $t^n$ as common denominator, this simplifies to $(p + qt)^n/(1-t)$, which is the SFGF of binomial distribution.

## 15. SFGF of geometric distribution

Obtain the SFGF of a geometric distribution with PMF $f(x; p)=q^x p, x = 0, 1, 2, \ldots, q = 1 - p$.

First find $P[X > n]$ to get the SF of a geometric distribution:

$$SF(n) = P(X > n) = q^{n+1}p + q^{n+2}p + \cdots . \qquad (3.19)$$

Take $q^{n+1}p$ as a common factor, and use $(1 + q + q^2 + \cdots) = 1/(1-q)$ to get the SF as $q^{n+1}p/(1-q) = q^{n+1}$. From this the SFGF follows as

$$SF_x(t; p) = q + q^2 t + q^3 t^2 + \cdots = q/(1 - qt) \quad \text{where } q = 1 - p. \qquad (3.20)$$

Thus, the SFGF is $q/(1 - qt)$. This follows directly by substituting $P_x(t) = p/(1 - qt)$ in $(1 - P_x(t))/(1-t)$ to get $[1 - p/(1 - qt)]/(1-t) = q(1-t)/[(1-t)(1-qt)] = q/(1 - qt)$. If the SF is defined as $P[X \ge k]$, this becomes $1/(1 - qt)$.

## 16. SFGF of Poisson distribution

Obtain the SFGF of a Poisson distribution with PMF $f(x; \lambda) = e^{-\lambda} \lambda^x / x!$, x=0,1,$\cdots$, $\infty$.

The PGF of Poisson distribution is found in Page 49 as $e^{-\lambda[1-t]}$. From this the CDFGF follows easily as $e^{-\lambda(1-t)}/(1-t)$. Write the SF as $SF = 1 - CDF$. Multiply both sides by $t^n$ and sum over the range to get

$$S_x(t) = 1/(1-t) - e^{-\lambda(1-t)}/(1-t) = \left[1 - e^{-\lambda(1-t)}\right]/(1-t). \qquad (3.21)$$

**17. SFGF of negative binomial distribution**

Obtain the SFGF of a negative binomial distribution with PMF $f(x; k, p) = \binom{x+k-1}{x}p^k q^x$, $x = 0, 1, 2, \ldots$.

The PGF of negative binomial distribution is derived in Page 50 as $[p/(1-qt)]^k$ from which the SFGF is obtained as $G(t) = [1 - p^k(1-qt)^{-k}](1-t)^{-1}$. As the SF $\Pr[x > c]$ of negative binomial distribution is $1 - I_q(c+1, k)$ for $k = 0, 1, 2, \ldots$ this can be used when $k$ is very large.

## 3.5  Generating Functions for Mean Deviation

It is shown in Sect. 3.3 that $P_x(t)/(1-t)$ is the CDFGF. A generating function for MD (MDGF) is found from CDFGF as follows. Write the CDFGF as

$$G_x(t) = \left(g_0 + g_1 t + g_2 t^2 + g_3 t^3 + \cdots\right), \qquad (3.22)$$

where $g_k$ denotes the sum of probabilities starting with the first, and $g_0 = p_0$. Divide both sides by $(1-t)$, and denote the LHS $G_x(t)/(1-t)$ by $D_x(t)$.

$$D_x(t) = \left(d_0 + d_1 t + d_2 t^2 + d_3 t^3 + \cdots\right), \qquad (3.23)$$

where $d_k = g_0 + g_1 + \cdots + g_k$, and $d_0 = g_0 = p_0$. The above step is valid using the same reasoning for CDFGF given above with $p_k = g_k$. This series also is absolutely convergent, as the sum of the probabilities are all less than or equal to one for any discrete distribution.

Now consider the MD of a discrete distribution with range [a,b]:

$$E|X - \mu| = \sum_{x=a}^{b} |x - \mu| p(x) \qquad (3.24)$$

where a is the lower and b is the upper limit of the distribution, and a could be $-\infty$ and b could be $+\infty$. As the mean of the random variable X is E(X)=$\mu$, it follows that E(X$-\mu$)=0, where E() is the expectation operator. Split the range of summation from a to $\lfloor \mu \rfloor$, and from $\lceil \mu \rceil$ to b:

$$E(X - \mu) = \sum_{x=a}^{\lfloor \mu \rfloor}(x - \mu)p(x) + \sum_{x=\lceil \mu \rceil}^{b}(x - \mu)p(x) = 0. \tag{3.25}$$

where $\lfloor \mu \rfloor$ denotes the floor operator (greatest integer less than $\mu$), and $\lceil \mu \rceil$ denotes the ceil operator (least integer greater than $\mu$). Note that $\lceil \mu \rceil = \lfloor \mu \rfloor + 1$. Write this as $\sum_{x=\lceil \mu \rceil}^{b}(x - \mu)p(x) = -\sum_{x=a}^{\lfloor \mu \rfloor}(x - \mu)p(x)$, and put in (3.25) to get

$$E|X - \mu| = \sum_{x=a}^{\lfloor \mu \rfloor}(\mu - x)p(x) - \sum_{x=a}^{\lfloor \mu \rfloor}(x - \mu)p(x) = 2\sum_{x=a}^{\lfloor \mu \rfloor}(\mu - x)p(x).$$

Apply the summation term-by-term inside the bracket to get

$$E|X - \mu| = 2\left(\mu F(\lfloor \mu \rfloor) - \sum_{x=a}^{\lfloor \mu \rfloor}xp(x)\right). \tag{3.26}$$

Sum the terms inside the bracket in reverse order of index variable as

$$\sum_{x=a}^{\lfloor \mu \rfloor}xp(x) = (\lfloor \mu \rfloor) * p(\lfloor \mu \rfloor) + (\lfloor \mu \rfloor - 1) * p(\lfloor \mu \rfloor - 1) + \cdots + a * p(a). \tag{3.27}$$

Collect alike terms on the RHS to get

$$\sum_{x=a}^{\lfloor \mu \rfloor}xp(x) = \mu * F(\lfloor \mu \rfloor) - \sum_{k=a}^{\lfloor \mu \rfloor}k * p(k), \tag{3.28}$$

where $F(\lfloor \mu \rfloor) = p(\lfloor \mu \rfloor) + p(\lfloor \mu \rfloor - 1) + \cdots + p(a)$. Now substitute in (3.26). The $\mu F(\lfloor \mu \rfloor)$ term cancels out, and we get

$$E|X - \mu| = 2\sum_{k=a}^{\lfloor \mu \rfloor}k\ p(k). \tag{3.29}$$

Write (3.29) as two summations

$$E|X - \mu| = 2\sum_{x=a}^{\lfloor \mu \rfloor}\sum_{i=a}^{x}p(i) \tag{3.30}$$

and substitute $\sum_{i=a}^{x}p(i) = F(x)$ to get

$$MD = 2\sum_{x=a}^{\lfloor \mu \rfloor}F(x), \tag{3.31}$$

where $\mu$ is the arithmetic mean, and $F(x)$ is the CDF. If the mean is not an integer, a correction term $2\delta F(c)$, where $\delta = \mu - c$, and $c = \lfloor \mu \rfloor$ must be added to get the correct MD. This correction term reduces to $F(\lfloor \mu \rfloor)$ when $\mu$ is a half-integer (e.g.: 3.5). Relating expression (3.31) as a coefficient in (3.23), we get the following theorem.

**Theorem 3.1** (Mean deviation generating function (MDGF)) *The MD of a discrete random variable is twice the coefficient of $t^{\lfloor \mu \rfloor - 1}$ in the power series expansion of $(1 - t)^{-2} P_x(t)$ where $\mu$ is the mean, $\lfloor \mu \rfloor$ denotes the integer part of $\mu$, and $P_x(t)$ is the PGF, with a correction term $2\delta F(\lfloor \mu \rfloor)$ added to it when $\mu$ is not an integer.*

In other words, the MDGF is $D_x(t) = 2P_x(t)/(1 - t)^2$.

### 3.5.1  Properties of MDGF

1. Particular cases: $D_x(0) = p_0$, $D'_x(0) = 2(p_0 + p_1)$.
2. $D_{x \pm c}(t) = t^{\pm c} D_x(t)$.
3. $D_x(t = -1) = \frac{1}{2} E[(-1)^x]$.
4. If X and Y are independent, $D_{x+y}(t) = D_x(t)P_y(t) = P_x(t)D_y(t)$.
5. If X and Y are independent, $D_{x-y}(t) = D_x(t)P_y(1/t)$.

---

## 3.6  MD of Some Distributions

This section derives the MD of some discrete distributions using above theorem. For each of the following distributions, a correction term $2\delta F(\lfloor \mu \rfloor)$ must be added when the mean is not an integer, where $\delta$ is the fractional part $\mu - \lfloor \mu \rfloor$, and $F()$ denotes the CDF. It then compares the results obtained using Eq. (3.12) and Theorem 3 to check if they tally.

### 3.6.1  MD of Geometric Distribution

The PGF of geometric distribution is $p/(1 - qt)$. Expanding $(1 - qt)^{-1}$ as an infinite series we get $P_x(t) = p(1 + qt + q^2t^2 + q^3t^3 + \cdots)$.

The CDFGF of a geometric distribution is obtained using $1 + q + q^2 + \cdots q^k = (1 - q^{k+1})/(1 - q) = (1 - q^{k+1})/p$ as

$$G(t) = \left(1 - q + t\left(1 - q^2\right) + t^2\left(1 - q^3\right) + t^3\left(1 - q^4\right) + \cdots\right) \tag{3.32}$$

because $p$ in the numerator cancels out with $1 - q = p$ in the denominator. Denote $(1 - q^{k+1})$ by $g_k$, and obtain the MDGF with coefficients $h_k = \sum g_k = \sum (1 - q^{k+1})$. As the mean of a geometric distribution is $q/p$, we can simply fetch the coefficient of $t^{\lfloor \mu \rfloor -1} = t^{\lfloor q/p \rfloor -1}$ in $H(t)$, and multiply by 2 to get the MD as $2 \sum_{k=0}^{\lfloor q/p \rfloor -1} (1 - q^{k+1})$, where $\lfloor q/p \rfloor$ denotes the integer part.

### 3.6.2  MD of Binomial Distribution

The PGF of binomial distribution $\text{BINO}(n, p)$ is $(q + pt)^n$. Hence, the MDGF is obtained as $2(q + pt)^n/(1 - t)^2$. The mean is $\mu = np$, and let $k = \lfloor np \rfloor$. Expand the denominator as an infinite series, and collect the coefficient of $t^{k-1}$ to get the MD as

$$MD = 2 \sum_{i=0}^{k} (k - i) \binom{n}{i} q^{n-i} p^i = 2npq \binom{n-1}{k} p^k q^{n-1-k}, \quad \text{where } k = \lfloor np \rfloor. \quad (3.33)$$

### 3.6.3  MD of Poisson Distribution

The PGF of Poisson distribution $P(\lambda)$ is $e^{\lambda(t-1)}$. According to Theorem 3, the MDGF is the coefficient of $t^{\lfloor \lambda \rfloor -1}$ in the power series expansion of $2e^{\lambda(t-1)}/(1 - t)^2$. Expanding the denominator as a power series and collecting the coefficient of $t^{\lfloor \lambda \rfloor -1}$ gives

$$MD = 2e^{-\lambda} \sum_{i=0}^{k} (k - i)\lambda^i /i! = 2e^{-\lambda}\lambda^{k+1}/k! \text{ where } k = \lfloor \lambda \rfloor. \quad (3.34)$$

### 3.6.4  MD of Negative Binomial Distribution

The PGF of negative binomial distribution $\binom{x+n-1}{n-1} p^n q^x$ is $(p/(1 - qt))^n$. Hence, the MDGF is obtained as $2p^n(1 - qt)^{-n}/(1 - t)^2$. The mean is $\mu = nq/p$, and let $k = \lfloor nq/p \rfloor$. Expanding the denominator as a power series and collecting the coefficient of $t^{\lfloor (nq/p) \rfloor -1}$ gives

$$MD = 2p^n \sum_{i=0}^{k} (k - i) \binom{i+n-1}{n-1} q^i \text{ where } k = \lfloor nq/p \rfloor. \quad (3.35)$$

### 3.6.5  MD of Logarithmic Distribution

The PGF of logarithmic distribution is $P_x(t, p) = \log(1 - qt)/\log(1 - q)$. The MDGF follows as $D_x(t) = 2\log(1 - qt)/[\log(1 - q)(1 - t)^2]$. When the mean is an integer, the MD is obtained as the coefficient of $t^{\mu-1}$ in the power series expansion of $D_x(t)$. As $\log(1 - q)$ is negative, we associate the negative sign with $\log(1 - qt)$ and expand it as an infinite series. The coefficient of $t^k$ in $\log(1 - qt)/(1 - t)^2$ is $kq + (k - 1)q^2/2 + (k - 2)q^3/3 + \cdots + 2q^{k-1}/(k - 1) + q^k/k$. As the mean is $\mu = q/[p \ \log(p)]$, it is a simple matter to extract the coefficient of $t^{\mu-1}$ and multiply it by $2/|\log(1 - q)|$ to get the MD.

---

## 3.7    Moment Generating Functions (MGF)

The MGF of a random variable is used to generate the moments algebraically. Let $X$ be a discrete random variable defined for all values of $x$. As $e^{tx}$ has an infinite expansion in powers of $x$ as $e^{tx} = 1 + (tx)/1! + (tx)^2/2! + \cdots + (tx)^n/n! + \cdots$, we multiply both sides by $f(x)$, and take expectation on both sides to get

$$M_x(t) = E(e^{tx}) = \begin{cases} \sum_x e^{tx} p(x) & \text{if } X \text{ is discrete;} \\ \int_{-\infty}^{\infty} e^{tx} f(x) dx & \text{if } X \text{ is continuous.} \end{cases}$$

In the discrete case this becomes

$$M_x(t) = \sum_{x=0}^{\infty} e^{tx} f(x) = 1 + \sum_{x=0}^{\infty} (tx)/1! f(x) + \sum_{x=0}^{\infty} (tx)^2/2! f(x) + \cdots . \qquad (3.36)$$

Replace each of the sums $\sum_{x=0}^{\infty} x^k f(x)$ by $\mu_k$ to obtain the following series (which is theoretically defined for all values, but depends on the distribution):

$$M_x(t) = 1 + \mu_1' t/1! + \mu_2' t^2/2! + \cdots + \mu_k' t^k/k! + \cdots . \qquad (3.37)$$

Analogous result holds for the continuous case by replacing summation by integration. By choosing $|t| < 1$, the above series can be made convergent for most random variables.

**Theorem 3.2** *The MGF (Sect. 3.7) and the PGF are connected as $M_x(t) = P_x(e^t)$, and $M_X(t = 0) = P_x(e^0) = P_x(1) = 1$.*

***Proof*** This follows trivially by replacing $t$ by $e^t$ in (3.5). Note that it is also applicable to continuous random variables. Put $t = 0$, and use $e^0 = 1$ to get the second part.  □

## 18. MGF of binomial distribution

If the PGF of BINO$(n, p)$ is $(q + pt)^n$, obtain the MGF and derive the mean.

The MGF can be found from Eq. (3.6) by replacing $t$ by $e^t$. This gives $M_x(t) = (q + pe^t)^n$. Take log to get $\log(M_x(t)) = n * \log(q + pe^t)$. Next, differentiate as above: $M'_X(t)/M_x(t) = n * pe^t/(q + pe^t)$. Put $t = 0$ to get the mean as $np$. Take log again to get $\log(M'_X(t)) - \log(M_x(t)) = \log(np) + t - \log(q + pe^t)$. Differentiate again, and denote $M'_X(t)$ simply by $M'$, etc. This gives $M''/M' - M'/M = 1 - pe^t/(q + pe^t)$. Put $t = 0$ throughout and use $M'(0) = np$ and $M(0) = 1$ to get $M''(0)/np - np = 1 - p$ or equivalently $M''(0) = (q + np) * np$. Finally, use $\sigma^2 = M''_X(0) - [M'_X(0)]^2 = (q + np) * np - (np)^2 = npq$.

### 3.7.1 Properties of Moment Generating Functions

1. MGF of an origin changed variate can be found from MGF of original variable

$$M_{x \pm b}(t) = e^{\pm bt} * M_x(t). \tag{3.38}$$

This follows trivially by writing $E[e^{t[x \pm b]}]$ as $e^{\pm bt} * E[e^{tx}]$.

2. MGF of a scale changed variate can be found from MGF of original variable as

$$M_{c*x}(t) = M_x(c * t). \tag{3.39}$$

This follows trivially by writing $E[e^{t\,cx}]$ as $E[e^{(ct)*x}]$.

3. MGF of origin-and-scale changed variate can be found from MGF of original variable as

$$M_{c*x \pm b}(t) = e^{\pm bt} * M_x(c * t). \tag{3.40}$$

This follows by combining both cases above.

**Theorem 3.3** (MGF of a sum) *The MGF of a sum of independent random variables is the product of their MGFs. Symbolically $M_{X+Y}(t) = M_x(t) * M_y(t)$.*

**Proof** We prove the result for the discrete case. $M_{X+Y}(t) = E(e^{t(x+y)}) = E(e^{tx}e^{ty})$. If $X$ and $Y$ are independent, we write the RHS as $\sum_x e^{tx} f(x) * \sum_y e^{ty} f(y) = M_x(t) * M_y(t)$. The proof for the continuous case follows similarly. This result can be extended to any number of pairwise independent random variables. □

If $X_1, X_2, \ldots, X_n$ are independent, and $Y = \sum_i X_i$ then $M_y(t) = \prod_i M_{X_i}(t)$.

### 19. Moments from $M_x(t)$

Prove that $E(X) = \frac{\partial}{\partial t} M_x(t)|_{t=0}$ and $E(X^2) = \frac{\partial^2}{\partial t^2} M_x(t)|_{t=0}$.

We know that $M_x(t) = E(e^{tx})$. Differentiating (3.37) w.r.t. $t$ gives $\frac{\partial}{\partial t} M_x(t) = \frac{\partial}{\partial t} E(e^{tx})$ $= E(\frac{\partial}{\partial t} e^{tx}) = E(xe^{tx})$ because $x$ is considered as a constant (and $t$ is our variable). Putting $t = 0$ on the RHS we get the result, as $e^0 = 1$. Differentiating a second time, we get $\frac{\partial^2}{\partial t^2} M_x(t) = \frac{\partial}{\partial t} E(xe^{tx}) = E(x^2 e^{tx})$. Putting $t = 0$ on the RHS we get $M_x''(t = 0) = E(x^2)$. Repeated application of this operation allows us to find the $k$th moment as $M_x^{(k)}(t = 0) = E(x^k)$. This gives $\sigma^2 = M_x''(t = 0) - [M_x'(t = 0)]^2$.

### 20. MGF of NBINO($k, p$) distribution

Find the MGF of negative binomial distribution NBINO($k, p$), and obtain the mean.

By definition,

$$M_x(t) = E(e^{tx}) = \sum_{x=0}^{\infty} \binom{x+k-1}{x} p^k q^x e^{tx} = \sum_{x=0}^{\infty} \binom{x+k-1}{x} p^k (qe^t)^x. \quad (3.41)$$

As $p^k$ is a constant, this is easily seen to be $[p/(1 - qe^t)]^k$. Take log and differentiate w.r.t. $t$, and put $t = 0$ to get $E(X) = kq/p$. By setting $p = 1/(1 + r)$ and $q = 1 - p = r/(1 + r)$, we get NBINO($k, 1/(1 + r)$) with PMF as $\binom{x+k-1}{x}[1/(1 + r)]^k [r/(1 + r)]^x$ with MGF $1/[1 + r(1 - e^t)]^k$.

### 21. MGF for central moments of Poisson distribution

Find the MGF for central moments of a Poisson distribution. Hence, show that $\mu_{r+1} = \lambda \left[ \binom{r}{1} \mu_{r-1} + \binom{r}{2} \mu_{r-2} + \cdots \binom{r}{r} \mu_0 \right]$.

First consider the ordinary MGF defined as $M_x(t) = E(e^{tx}) = \sum_{k=0}^{\infty} e^{tx} e^{-\lambda} \lambda^x / x! = e^{-\lambda} \sum_{k=0}^{\infty} (\lambda e^t)^x / x! = e^{-\lambda} e^{\lambda e^t} = e^{\lambda e^t - \lambda} = e^{\lambda(e^t - 1)}$. As the mean of a Poisson distribution is $\mu = \lambda$, we use the Property 1. to get

$$M_{x-\lambda}(t) = e^{-\lambda t} M_x(t) = e^{-\lambda t} e^{\lambda(e^t - 1)} = e^{\lambda(e^t - t - 1)} = \sum_{j=0}^{\infty} \mu_j t^j / j!. \quad (3.42)$$

Differentiate $\sum_{j=0}^{\infty} \mu_j t^j / j! = e^{\lambda(e^t - t - 1)}$ by $t$ to get

$$\sum_{j=0}^{\infty} \mu_j j t^{j-1} / j! = e^{\lambda(e^t - t - 1)} \lambda \left( e^t - 1 \right) = \lambda \left( e^t - 1 \right) \sum_{j=0}^{\infty} \mu_j t^j / j!. \quad (3.43)$$

Expand $e^t$ as an infinite series. The RHS becomes $\lambda \sum_{k=1}^{\infty} t^k / k! \sum_{j=0}^{\infty} \mu_j t^j / j!$. This can be written as $\lambda \sum_{k=1}^{\infty} \sum_{j=0}^{\infty} \mu_j t^{j+k} / [j! k!]$. Equate coefficients of $t^r$ on both sides to get

$\mu_{r+1}/r! = \lambda \left[ \frac{\mu_{r-1}}{1!(r-1)!} + \frac{\mu_{r-2}}{2!(r-2)!} + \cdots + \frac{\mu_0}{r!(r-r)!} \right]$. Cross-multiplying and identifying the binomial coefficients, this becomes

$$\mu_{r+1} = \lambda \left[ \binom{r}{1} \mu_{r-1} + \binom{r}{2} \mu_{r-2} + \cdots + \binom{r}{r} \mu_0 \right]. \tag{3.44}$$

## 22. MGF of power-series distribution

Find the MGF of power-series distribution.

The PMF of power-series distribution is $P(X = k) = a_k \theta^k / F(\theta)$ where $k$ takes any integer values. Then $M_x(t) = E(e^{tx}) = \sum_{k=0}^{\infty} e^{tx} a_k \theta^k / F(\theta)$.

**Theorem 3.4** *Prove that if $E[|X|^k]$ exists and is finite, then $E[|X|^j]$ exists and is finite for each $j < k$.*

***Proof*** We prove the result for the continuous case. The proof for discrete case follows easily by replacing integration by summation. As $E[|X|^k]$ exists, we have $\int_x |X|^k dF(x) < \infty$. Now consider an arbitrary $j < k$ for which

$$\int_{-\infty}^{\infty} |x|^j dF(x) = \int_{-1}^{+1} |x|^j dF(x) + \int_{|x|>1} |x|^j dF(x). \tag{3.45}$$

As $j < k$, $|x|^j < |x|^k$ for $|x| > 1$. Hence integral (3.45) becomes

$$\int_{-\infty}^{\infty} |x|^j dF(x) < \int_{-1}^{+1} |x|^j dF(x) + \int_{|x|>1} |x|^k dF(x)$$

$$\leq \int_{-1}^{+1} dF(x) + \int_{|x|>1} |x|^k dF(x). \tag{3.46}$$

The RHS of (3.46) is upper bounded by $1 + E[|X|^k]$, and is $< \infty$. This proves that the LHS exists for each $j$. □

## 3.8    Characteristic Functions

The MGF of a distribution may not always exist. Those cases can be dealt with in the complex domain by finding the expected value of $e^{itx}$, where $i = \sqrt{-1}$, which always exist. Thus, the ChF of a random variable is defined as

$$\text{ChF} = E(e^{itx}) = \begin{cases} \sum_{x=-\infty}^{\infty} e^{itx} p_x & \text{if } X \text{ is discrete;} \\ \int_{x=-\infty}^{\infty} e^{itx} f(x)dx & \text{if } X \text{ is continuous.} \end{cases}$$

We have seen above that the ChF, if it exists, can generate the moments. Irrespective of whether the random variable is discrete or continuous, we could expand the ChF as a McClaurin series as

$$\phi_x(t) = \sum_{j=0}^{\infty} \mu'_j (it)^j / j! = \phi(0) + t\phi'(0) + (t^2/2!)\phi''(0) + \cdots, \tag{3.47}$$

which is convergent for an appropriate choice of $t$ (which depends on the distribution). As $\phi_x(t)$ in the continuous case can be represented as $\phi_x(t) = \int_{-\infty}^{\infty} e^{itx} dF(x)$, successive derivatives w.r.t. $t$ gives $\int i^n x^n dF(x) = i^n \mu'_n$. Define $\delta^{(n)}(x)$ as the $n$th derivative of the delta function. Then the PMF can be written as an infinite sum as

$$f(x) = \sum_{j=0}^{\infty} (-1)^j \mu'_j \delta^{(j)}(x)/j!. \tag{3.48}$$

Characteristic functions are used to study various properties of a random variable analytically.

### 23. ChF of Poisson distribution

Find the ChF of the Poisson distribution.

Consider $\phi_x(t) = E(e^{itx}) = \sum_{k=0}^{\infty} e^{itx} e^{-\lambda} \lambda^x / x!$. Take $e^{-\lambda}$ outside the summation to get $e^{-\lambda} \sum_{k=0}^{\infty} (\lambda e^{it})^x / x! = e^{-\lambda} e^{\lambda e^{it}} = e^{\lambda e^{it} - \lambda} = e^{\lambda(e^{it} - 1)}$.

## 3.8.1  Properties of Characteristic Functions

Characteristic functions are Laplace transforms of the corresponding PMF. As all Laplace transforms have an inverse, we could invert it to get the PMF. Hence, there is a one-to-one correspondence between the ChF and PMF. This is especially useful for continuous distributions as shown below (Table 3.2). There are many simple properties satisfied by the ChF.

1. $\overline{\phi(t)} = \phi(-t)$, $\phi(0) = 1$, and $|\phi(\pm t)| \leq 1$. In words, this means that the complex conjugate of the ChF is the same as that obtained by replacing $t$ with $-t$ in the ChF. The assertion $\phi(0) = 1$ follows easily because this makes $e^{itx}$ to be 1.
2. $\phi_{ax+b}(t) = e^{ibt} \phi_x(at)$. This result is trivial as it follows directly from the definition.

**Table 3.2** Table of characteristic functions

| Distribution | Density function | Characteristic function |
|---|---|---|
| Bernoulli | $p^x(1-p)^{1-x}$ | $q + pe^{it}$ |
| Binomial | $\binom{n}{x}p^xq^{n-x}$ | $(q + pe^{it})^n$ |
| Negative bino. | $\binom{x+k-1}{x}p^kq^x$ | $p^k(1 - qe^{it})^{-k}$ |
| Poisson | $e^{-\lambda}\lambda^x/x!$ | $\exp(\lambda(e^{it} - 1))$ |
| Rectangular | $f(x) = \Pr[X = k] = 1/N$ | $(1 - e^{itN})/[N(e^{-it} - 1)]$ |
| Geometric | $q^xp$ | $p/(1 - qe^{it})$ |
| Logarithmic | $q^x/[-x \log p]$ | $\ln(1 - qe^{it})/\ln(1 - q)$ |
| Multinomial | $(n!/\prod_{i=1}^{k} x_i!)*\prod_{i=1}^{k} p_i^{x_i}$ | $[\sum_{j=1}^{k} p_j e^{it_j}]^n$ |

3. If $X$ and $Y$ are independent, $\phi_{ax+by}(t) = \phi_x(at) * \phi_y(bt)$. Putting $a = b = 1$, we get $\phi_{x+y}(t) = \phi_x(t) * \phi_y(t)$ if $X$ and $Y$ are independent.
4. $\phi_x(t)$ is continuous in $t$, and convex for $t > 0$. This means that if $t_1$ and $t_2$ are two values of $t > 0$, then $\phi_x((t_1 + t_2)/2) \leq \frac{1}{2}[\phi_x(t_1) + \phi_x(t_2)]$.
5. $\partial^n \phi_x(t)/\partial t^n \mid_{t=0} = i^n E(X^n)$.

### 24. Property of ChF of symmetric random variables

Prove that the random variable $X$ is symmetric about the origin iff the ChF $\phi_x(it)$ is real-valued for all $t$.

Assume that $X$ is symmetric about the origin, so that $f(-x) = f(x)$. Then for a bounded and odd Borel function $g(x)$ we have $\int g(x)dF(x) = 0$. As $g(x)$ is odd, this is equivalent to $\int \sin(tx)dF(x) = 0$. Hence, $\phi_x(t) = E(e^{itx}) = E[\cos(tx)]$ is real. Also as $\phi_{-x}(t) = \phi_x(-t) = \overline{\phi}_x(t) = \phi_x(\bar{t})$, $F_X(x)$ and $F_{-X}(x)$ are the same.

**Remark 3.1** The characteristic function uniquely determines a distribution.

The inversion theorem provides a means to find the PMF from the characteristic function as $f(x) = 1/(2\pi) \int_{-\infty}^{\infty} \phi_x(it)e^{-itx} dt$.

### Uniqueness Theorem

Let random variables $X$ and $Y$ have MGF $M_x(t)$ and $M_y(t)$, respectively. If $M_x(t) = M_y(t) \forall t$, then $X$ and $Y$ have the same probability distribution. This is very similar to the result for PGFs.

## 3.9   Cumulant Generating Functions

The CGF is slightly easier to work with for exponential, normal and Poisson distributions. It is defined in terms of the MGF as $K_x(t) = \ln(M_x(t)) = \sum_{j=1}^{\infty} k_j t^j / j!$, where $k_j$ is the $j$th cumulant. This relationship shows that cumulants are polynomial functions of moments (low order cumulants can also be exactly equal to the corresponding moments). For example, for the general univariate normal distribution with mean $\mu_1 = \mu$ and variance $\mu_2 = \sigma^2$, the first and second cumulants are, respectively, $\kappa_1 = \mu$ and $\kappa_2 = \sigma^2$.

**Theorem 3.5**  *Prove* $K_{aX+b}(t) = bt + K_x(at)$.

**Proof**  $K_{aX+b}(t) = \log(M_{aX+b}(t)) = \log(e^{bt} M_X(at)) = bt + \log(M_X(at)) = bt + K_x(at)$ using $\log(ab) = \log(a) + \log(b)$, and $\log(e^x) = x$.                                        □

**Theorem 3.6**  *The CGF of an origin and scale changed variable is* $K_{(X-\mu)/\sigma}(t) = (-\mu/\sigma)t + K_x(t/\sigma)$.

**Proof**  This follows from the above theorem by setting $a = 1/\sigma$ and $b = -\mu/\sigma$. The cumulants can be obtained from moments and *vice versa*. This holds for cumulants about any origin (including zero) in terms of moments about the same origin.                    □

### 3.9.1   Relations Among Moments and Cumulants

The central and raw moments are related as $\mu_k = E(X - \mu)^k = \sum_{j=0}^{k} \binom{k}{j}(-\mu)^{k-j} \mu_j'$ (Chattamvelli and Shanmugam 2020). As the CGF of some distributions are easier to work with, we can find cumulants and use the relationship with moments to obtain the desired moment.

**Theorem 3.7**  *The $r$th cumulant can be obtained from the CGF as* $\kappa_r = \frac{\partial^r K_x(t)}{\partial t^r}\big|_{t=0}$.

**Proof**  We have

$$K_x(t) = \sum_{r=0}^{\infty} \kappa_r t^r / r! = \kappa_0 + \kappa_1 t + \kappa_2 t^2/2! + \cdots . \tag{3.49}$$

As done in the case of MGF, differentiate (3.49) $k$ times and put $t = 0$ to get the $k$th cumulant.                                        □

## 25. Moments from cumulants

Prove that $\kappa_1 = \mu_1, \kappa_2 = \mu_2 = \sigma^2$, and $\kappa_3 = \mu_3 = E(X - \mu)^3$.

We know that $K_x(t) = \log(M_x(t))$ or equivalently $M_x(t) = \exp(K_x(t))$. We expand $M_x(t) = 1 + t/1!\mu_1 + t^2/2!\mu_2' + t^3/3!\mu_3' + \cdots$ and substitute for $K_x(t)$ also to get

$$\sum_{r=0}^{\infty} \mu_r' t^r/r! = \exp\left(\sum_{r=0}^{\infty} \kappa_r t^r/r!\right). \tag{3.50}$$

Differentiate $n$ times, and put $t = 0$ to get

$$\mu_{n+1}' = \sum_{j=0}^{n} \binom{n}{j} \mu_{n-j}' \kappa_{j+1}. \tag{3.51}$$

Put $n = 0, 1, \ldots$ to get the desired result. There is another way to get the low-order cumulants. Truncate $M_x(t)$ as $1 + t/1!\mu_1 + t^2/2!\mu_2' + t^3/3!\mu_3'$. Expand the RHS using $\log(1 + x) = x - x^2/2 + x^3/3 - x^4/4 \ldots$ where $x = (t/1!)\,\mu_1 + (t^2/2!)\,\mu_2' + (t^3/3!)\,\mu_3'$, and collect similar terms to get

$$K_x(t) = \mu_1 t + (\mu_2' - \mu_1^2)t^2/2! + (\mu_3' - 3\mu_1\mu_2' + 2\mu_1^3)t^3/3! + \cdots. \tag{3.52}$$

Compare the coefficients of $t^k/k!$ to get $\kappa_1 = \mu_1, \kappa_2 = \mu_2 - \mu_1^2 = \sigma^2, \kappa_3 = (\mu_3' - 3\mu_1\mu_2' + 2\mu_1^3) = E(X - \mu)^3$.

Next write $M_x(t)$ as $1 + [tx/1! + (tx)^2/2! + \cdots]$, expand $K_x(t) = \log(M_x(t))$ as an infinite series to get

$$\sum_{r=0}^{\infty} \kappa_r t^r/r! = [tx/1! + (tx)^2/2! + \cdots] - [tx/1! + (tx)^2/2! + \cdots]^2/2! + \cdots. \tag{3.53}$$

Equate like coefficients of $t$ to get

$$\kappa_{n+1} = \mu_{n+1}' - \sum_{j=0}^{n-1} \binom{n}{j} \mu_{n-j}' \kappa_{j+1}. \tag{3.54}$$

## 3.10 Factorial Moment Generating Functions

There are two types of factorial moments known as *falling factorial* and *rising factorial* moments. Among these, the falling factorial moments are more popular. The $k$th (falling) factorial moment of $X$ is defined as $E[X(X - 1)(X - 2) \cdots (X - k + 1)] = E[X!/(X - k)!]$, where $k$ is an integer $\geq 1$. It is easier to evaluate for those distributions that have an $x!$ or $\Gamma(x + 1)$ in the denominator (e.g., binomial, negative binomial, hypergeometric, and

Poisson distributions). The factorial moments and ordinary moments are related through the Stirling number of the first kind as follows:

$$X!/(X-r)! = \sum_{j=0}^{r} s(r, j)X^{j} \Rightarrow \mu'_{(r)} = \sum_{j=0}^{r} s(r, j)\mu'_{j}. \tag{3.55}$$

A reverse relationship exists between ordinary and factorial moments using the identity $X^{r} = \sum_{j=0}^{r} S(r, j)X!/(X-j)!$ as $\mu'_{r} = \sum_{j=0}^{r} S(r, j)\mu'_{(j)}$ where $S(r, j)$ is the Stirling number of second kind. FCGF can analogously be defined as $FK_{x}(t) = \ln(FM_{x}(t))$, where $FM_{x}(t)$ is the FMGF, and ln denotes log to the base $e$.

There are two ways to get (falling) factorial moments. The simplest way is by differentiating the PGF (see Sect. 3.2.1). As $P_{x}(t) = E(t^{x}) = E(e^{x\log(t)}) = M_{X}(\log(t))$, we could differentiate it $k$ times and put $t = 1$ to get factorial moments.

We define it as $\Gamma_{x}(t) = E\left[(1+t)^{x}\right]$, because if we expand it using binomial theorem we get

$$E\left[(1+t)^{x}\right] = E\left[1 + tx + t^{2}x(x-1)/2! + t^{3}x(x-1)(x-2)/3! + \cdots\right]. \tag{3.56}$$

Obviously $\Gamma_{x}(t)|_{t=1} = 1$, $\Gamma'_{x}(t)|_{t=1} = E(X) = \mu$. Factorial moments are obtained by taking term by term expectations on the RHS.

The rising factorial moment is defined as $E[X(X+1)(X+2)\ldots(X+k-1)] = E[(X+k-1)!/(X-1)!]$. An analogous expression can also be obtained for rising factorials using the expansion $(1-t)^{-x} = \sum_{k=0}^{\infty} \binom{k+x-1}{k}t^{k}$. Taking term-by-term expectations as $E\left[(1-t)^{-x}\right] = E\left[1 + tx + t^{2}x(x+1)/2! + t^{3}x(x+1)(x+2)/3! + \cdots\right]$, we get rising FMGF. We could also get rising factorial moments from $P_{-x}(t) = E(t^{-x})$. Differentiating it once gives $P'_{-x}(t) = E\left(-xt^{-x-1}\right)$. From this we get $P'_{-x}(1) = E(-x)$. Differentiate it $r$ times to get $P^{(r)}_{-x}(t) = E\left(-x(-x-1)(-x-2)\ldots(-x-r+1)t^{-x-r}\right)$. Putting $t = 1$, this becomes

$$P^{(r)}_{x}(1) = (-1)^{r} E\left[x(x+1)\ldots E(x+r-1)\right] = (-1)^{r}\mu^{(r)}. \tag{3.57}$$

Both (rising and falling) FMGFs are of OGF type. Replacing the summation by integration gives the corresponding results for the continuous distributions.

To distinguish between the two, we will denote the falling factorial moment as $E(X_{(k)})$ or $\mu_{(k)}$, and the rising factorial moment as $E(X^{(k)})$ or $\mu^{(k)}$. Unless otherwise specified, factorial moment will mean falling factorial moment $\mu_{(k)}$.

## 26. Factorial moments of the Poisson distribution

Find the $k$th factorial moments of the Poisson distribution, and obtain the first two moments.
  By definition,

$$\mu_{(k)} = \sum_{x=0}^{\infty} x(x-1)(x-2)\ldots(x-k+1)e^{-\lambda}\lambda^x/x!$$

$$= e^{-\lambda}\lambda^k \sum_{x=k}^{\infty} \lambda^{x-k}/(x-k)!$$

$$= e^{-\lambda}\lambda^k \sum_{y=0}^{\infty} \lambda^y/y! = e^{-\lambda}\lambda^k e^{\lambda} = \lambda^k. \tag{3.58}$$

Another way is using the MGF. Write the MGF as $\exp(-\lambda)\exp(\lambda t)$. The $k$th derivative is easily found as $\exp(-\lambda)\lambda^k \exp(\lambda t) = \lambda^k \exp(-\lambda(1-t))$. Now put $t = 1$ to get the result. Alternatively, we could obtain the FMGF directly, and get the desired moments:

$$\text{FMGF}_x(t) = E[(1+t)^x] = \sum_{x=0}^{\infty}(1+t)^x e^{-\lambda}\lambda^x/x! = e^{-\lambda}\sum_{x=0}^{\infty}[\lambda(1+t)]^x/x! = e^{\lambda t}.$$
$$\tag{3.59}$$

The $k$th factorial moment is obtained by differentiating this expression $k$ times and putting $t = 0$. We know that the $k$th derivative of $e^{\lambda t}$ is $\lambda^k e^{\lambda t}$, from which $k$th factorial moment is obtained as $\lambda^k$. Putting $k = 1$ and 2 gives the desired moments.

**Theorem 3.8** *The factorial moments of a Poisson distribution are related as* $\mu_{(k)} = \lambda^r \mu_{(k-r)}$. *In particular* $\mu_{(k)} = \lambda\mu_{(k-1)}$.

***Proof*** This follows easily because $\mu_{(k)}/\mu_{(k-r)} = \lambda^k/\lambda^{k-r} = \lambda^r$.                          □

## 27. FMGF of the geometric distribution

Find the FMGF of the geometric distribution.
  By definition, FMGF is:

$$E\left[(1+t)^x\right] = \sum_{x=0}^{\infty}(1+t)^x q^x p = p\sum_{x=0}^{\infty}[q(1+t)]^x = p/[1-q(1+t)]. \tag{3.60}$$

As $1 - q = p$, the denominator becomes $[1 - q(1+t)] = p - qt$. Hence FMGF = p/(p-qt). The $k$th factorial moment is obtained by differentiating this expression $k$ times and putting

$t = 0$. We know that the $k$th derivative of $1/(ax + b)$ is $a^r r!(-1)^r/(ax + b)^{r+1}$. Hence, $k$th derivative of $1/[1 - q(1 + t)]$ is $r!q^r/p^{r+1}$, as $q = 1 - p$. This gives the $k$th factorial moment as $pr!q^r/p^{r+1} = r!(q/p)^r$. Alternatively, find $E(t^x)$, differentiate $k$ times and put $t = 1$ to get the same result.

### 28. FMGF of the negative binomial distribution

Find the FMGF of the negative binomial distribution.

We have seen in Sect. 3.2 that the PGF of negative binomial distribution is $(p/(1 - qt))^k$. As the PGF and FMGF are related as $FMGF(t) = PGF(1 + t)$, we get the FMGF as $(p/(1 - q(1 + t)))^k$. As $1 - q = p$, this can be written as $(p/(p - qt))^k$ or $(1 - qt/p)^{-k}$.

## 3.11   Conditional Moment Generating Functions (CMGF)

Consider an integer random variable that takes values $\geq 1$. We define a sum of independent random variables as $S_N = \sum_{i=1}^{N} X_i$. For a fixed value of $N = n$, the distribution of the finite sum $S_n$ can be obtained in closed form for many distributions when the variates are independent. The conditional MGF can be expressed as

$$M_{X|Y}(t) = \sum_x f(x|y)e^{tx}. \tag{3.61}$$

Replacing $t$ by "$it$" gives the corresponding conditional characteristic function. If the variates are mutually independent, this becomes $M_{S|N}(t|N) = [M_x(t)]^N$.

### 29. Compound Poisson distribution

Find the MGF of the compound Poisson distribution $Y = \sum_{k=1}^{N} X_k$ where $N$ has Poisson distribution with mean $\lambda$ and each $X_k$'s have Poisson distribution with mean $\mu_k$.

The PMF of $Y$ is $\Pr(Y = r) = \sum_{j=0}^{\infty} P(Y|N)P(N = j)$. After simplification this becomes $G(t) = \exp(-\lambda[1 - \exp(-\mu(1 - t))])$. Replace $t$ by $\exp(t)$ to get the MGF.

## 3.12   Generating Functions of Truncated Distributions

The GFs of truncated distributions can easily be found from that of non-truncated ones. As an example, the classical binomial distribution has PGF $(q + pt)^n$ and MGF $(q + pe^t)^n$. The corresponding PGF and MGF of zero-truncated distribution are, respectively, $(q + pt)^n/(1 - q^n)$ and $(q + pe^t)^n/(1 - q^n)$. In general, the PGF of a zero-truncated distribution can be expressed as $P_z(t) = (P(t) - P(0))/[1 - P(0)]$ where $P(0)$ is the prob-

**Table 3.3** Summary table of zero-truncated generating functions

| Distribution | PGF | z-PGF | z-MGF |
|---|---|---|---|
| Binomial | $(q + pt)^n$ | $(q + pt)^n/(1 - q^n)$ | $(q + pe^t)^n/(1 - q^n)$ |
| Poisson | $\exp(-\lambda[1\text{-}t])$ | $e^{(-\lambda[1-t])}/[1 - e^{-\lambda}]$ | $e^{-\lambda[1-e^t]}/[1 - e^{-\lambda}]$ |
| Geometric | p/(1-qt) | p/[q(1-qt)] | p/[q(1-q exp(t))] |
| Neg. Binom. | $[p/(1\text{-}qt)]^k$ | $[p/(1\text{-}qt)]^k/(1-p^k)$ | $1/(1-p^k)[p/(1\text{-}q\ \exp(t))]^k$ |

z-PGF is zero-truncated PGF. Higher order truncations will bring additional terms in the denominator

ability that $x = 0$. The PGF of a binomial distribution truncated at $k$ is $(q + pt)^n/(1 - \sum_{j=0}^{k} \binom{n}{j} p^j q^{n-j})$. The $k$-truncated MGF is obtained by replacing each $t$ with $\exp(t)$. A similar expression results for other distributions. Thus, a $k$-truncated Poisson distribution has PGF $\exp(-\lambda[1 - t])/\left[1 - \sum_{j=0}^{k} \exp(-\lambda)\lambda^j/j!\right]$ (Table 3.3).

## 3.13 Convergence of Generating Functions

Properties of GFs are useful in deriving the distributions to which a sequence of GFs converge. Let $X_n$ be a sequence of random variables with ChF $\phi_X^n(t)$. If $\lim_{n\to\infty} \phi_X^n(t)$ converges to a unique limit, say $\phi_x(t)$ for all points in a neighborhood of $t = 0$, then that limit determines the unique CDF to which the distribution of $X_n$ converge. Symbolically,

$$\lim_{n\to\infty} \phi_X^n(t) = \phi_x(t) \Rightarrow \lim_{n\to\infty} F_{X_i}(x) = F(x). \tag{3.62}$$

## 3.14 Summary

This chapter introduced various GFs encountered in statistics. These have wide applications in many other fields including astrophysics, fluid mechanics, spectroscopy, polymer chemistry, bio-informatics, and various engineering fields. Examples are included to illustrate the use of various GFs. The classical PGFs of discrete distributions are extended to get CDFGF, SFGF, and MDGF. This is then used to find the mean deviation by extracting just one coefficient $[t^{\lfloor \mu \rfloor - 1}]2P_x(t)/(1 - t)^2$, where $\mu$ is the mean.

# References

Chattamvelli, R., & Shanmugam, R. (2020). *Discrete distributions in engineering and the applied sciences*. Springer.

Johnson, N. L., Kotz, S., & Balakrishnan, N. (2005). *Continuous univariate distributions* (Vols. 1,2, 2nd ed.). New York: Wiley.

Puri, P. S. (1966). Probability generating functions of absolute difference of two random variables. *Proceedings of the National Academy of Sciences, 56*(4), 1059–1061.

# Applications of Generating Functions

**4**

This chapter discusses applications of GF in various fields. In mathematics, it is mainly used in algebra, geometry, partial differential equations, number theory, graph theory and combinatorics. Applications areas in chemistry include physical chemistry and organic chemistry (especially polymer chemistry). Computer science applications include analysis of algorithms, formal languages and data structures. Solution of linear recurrence relations are discussed and its application to Towers of Hanoi puzzle is explored. Most common applications in bioinformatics and genomics are also briefly discussed. This is followed by applications in civil engineering (especially structural engineering) and reliability. The chapter ends with some applications in statistics, epidemiology and management.

## 4.1  Applications in Algebra

There are many applications of GFs in algebra. GFs can be used to find the number of solutions to a single linear equation. Consider a simple example of an equation $x + y + z = r$, where $x, y, z$ are non-negative integers, and $r$ is a constant. Consider the OGF $(1 + t + t^2 + t^3 + \cdots + t^r)$. Then there is a one-to-one correspondence between the number of solutions to the above equation, and the coefficient of $t^r$ in the power-series expansion of $(1 + t + t^2 + t^3 + \cdots + t^r)^3$. This can be greatly simplified if further restrictions are placed on the possible values that $x$, $y$, and $z$ can take. As an example, if each variable is restricted to be an integer in the range $0 \leq \{x, y, z\} \leq c$, our OGF reduces to $(1 + t + t^2 + t^3 + \cdots + t^c)^3$. If each of them is restricted to be real numbers in $[0, 1]$ range, this becomes $[t^r](1 + t)^3$. In general, if we have an equation $x_1 + x_2 + \cdots + x_n = r$ where the $x_i'$s are in $[0, 1]$, the number of solutions is $[t^r](1 + t)^n$ which is $\binom{n}{r}$. If the number of

R. Chattamvelli and R. Shanmugam, *Generating Functions in Engineering and the Applied Sciences*, Synthesis Lectures on Engineering, Science, and Technology, https://doi.org/10.1007/978-3-031-21143-0_4

non-negative integer solutions are sought, we get the OGF $(1 + t + t^2 + t^3 + \cdots + t^r)^n = \sum_{r=0}^{\infty} \binom{n+r-1}{r} t^r$. Similar approach holds for other suitable restrictions.

Next consider a constrained linear equation of the form $x + y + z = r$ where $x$ is even integer, $y$ is $\geq 0$ and $0 \leq z \leq 1$. As $x$ is allowed to be only even integers, it has OGF $1 + t^2 + t^4 + \cdots = 1/(1 - t^2)$. As $y$ can take any nonnegative integer value, it has OGF $1/(1 - t)$. As $z$ can take values 0 or 1 only, its OGF is $(1 + t)$. Hence, the required number is the coefficient of $t^r$ in $(1/(1 - t^2))(1/(1 - t))(1 + t)$. Write $(1 - t^2)) = (1 + t)(1 - t)$ and cancel $(1 + t)$ from numerator and denominator, to get $F(t) = 1/(1 - t)^2$. The coefficient of $t^r$ is $(r + 1)$.

### 4.1.1   Series Involving Integer Parts

The integer part or "floor" of a decimal number is the greatest integer less than or equal to that number. This is denoted by the floor-operator $\lfloor x \rfloor$ in computing, and allied fields. Thus, $\lfloor 3.14 \rfloor = \lfloor 3 \rfloor = 3$. The least integer greater than or equal to $x$ is called the "ceil," and denoted by $\lceil x \rceil$ operator. Thus, $\lceil 3.14 \rceil = 4$. When $x$ is an integer, the ceil and floor operators return $x$ itself ($\lceil x \rceil = \lfloor x \rfloor = x$). Otherwise, these satisfy $\lceil x \rceil - \lfloor x \rfloor = 1$, $\lceil -x \rceil = -\lfloor x \rfloor$, $-\lfloor x \rfloor = -\lceil x \rceil$. The fractional part is $(x - \lfloor x \rfloor)$ when $x$ is not an integer. Thus, we have $\lfloor x + \frac{1}{2} \rfloor + \lfloor x \rfloor = \lfloor 2x \rfloor$.

Series involving integer parts occurs in many problems. Consider a tournament (like tennis) with $n$ teams. If $n$ is even, we could evenly divide the teams into two groups of $n/2$ each and identify the winner in each game. Those who win in the first round are again paired for the second round, and the process goes on until one final round decides who wins the game. Thus, the total number of games played is $\lfloor (n + 1) \rfloor / 2 + \lfloor (n + 1) \rfloor / 2^2 + \lfloor (n + 1) \rfloor / 2^3 + \cdots + 1$. A team (persons) that does not have a matching-pair when there are an odd number of teams will play with one of the winners, thereby reducing the number of teams by one. For example, if there are 7 teams, there will be $(3 + 1) = 4$ games in the first tournament, 2 games in the second and a final. Several applications in graph theory also use these operators. For instance, the number of edge-crossings of a complete bipartite graph $K_{m,n}$ with $m$ vertices in the first set, and $n$ vertices in the other is given by $\lfloor n/2 \rfloor \lfloor (n - 1)/2 \rfloor \lfloor m/2 \rfloor \lfloor (m - 1)/2 \rfloor$.

### 1. Sum of series involving integer parts

Find the OGF of $\lfloor (n + 1) \rfloor / 2$, $\lfloor (n + 2) \rfloor / 2^2$, $\lfloor (n + 2^2) \rfloor / 2^3$, ..., $\lfloor (n + 2^k) \rfloor / 2^{k+1}$, ..., and evaluate the sum.

Write $F(t) = \sum_{k=0}^{\infty} \lfloor (n + 2^k)/2^{k+1} \rfloor t^k$ as $\sum_{k=0}^{\infty} \lfloor (n/2^{k+1} + 1/2) \rfloor t^k$. Now use $\lfloor x + \frac{1}{2} \rfloor = \lfloor 2x \rfloor - \lfloor x \rfloor$ to get $F(t) = \sum_{k=0}^{\infty} \lfloor n/2^k \rfloor t^k - \sum_{k=0}^{\infty} \lfloor n/2^{k+1} \rfloor t^k$. The coefficients form a telescopic series (Chap. 1) $\sum_{k=0}^{\infty} [\lfloor n/2^k \rfloor - \lfloor n/2^{k+1} \rfloor] t^k$. Put t=1 so that the alternate terms cancel out giving $\lfloor n \rfloor$ as the answer.

## 4.1.2  Permutation Inversions

Collaborative filtering is a technique used by several e-commerce sites to rank the preferences of several customers, and filter out maximum matches among a set of customers. This is then used to suggest potential purchases (unselected items) that a customer may be interested in, by using the purchases made by other similar customers. Thus, a person who buys a programming book gets a list of similar books purchased in the past by other customers who also bought that book. Similarly, some e-com sites have a product rating option in which a customer who purchased an item can give a score (say 5-star to 1-star) using which the site recommends other similar items that people who have rated it compatibly have liked or purchased. The items can be anything like food, electronics, books, movies, music, etc. But the following discussion assumes that there are $n$ items to be ranked from the same domain. If thousands of rankings are already available, we could come up with an order based on the frequency as $(a_1 < a_2 < \cdots < a_n)$ where $a_1$ is the item that got the least vote or preference, and $a_n$ is the item with greatest vote.

Consider a set of $n$ labeled distinct elements that can be ordered among themselves using a relational operator. There are $n!$ permutations of $n$ elements. This means that we could use numerical comparison when elements are real numbers, modulus comparison for complex numbers, alphabetic comparison when elements are alphabets or characters, string comparison for strings, distance comparison for geometric objects and so on. A permutation of the elements is an arrangement of them in a particular order. Suppose we compare all possible pairs of elements except $(k, k)$. As the first element can be compared with $(n - 1)$ other elements, second with $(n - 2)$ other elements and so on, the penultimate element can be compared with the last one, we have a total of $1 + 2 + \cdots + (n - 1)$ set of pairs $(j, k)$ where $j \neq k$. This simplifies to $m = n(n - 1)/2 = \binom{n}{2}$. Out of the m pairs, we call a pair $(j, k)$ as an *inversion* if the corresponding elements are out-of-order (i.e., the ordering operator returns a FALSE result).

The concept of inversion is also used in many other fields like computational biology and genomics, where two DNA sequences are said to be in inversion when there is an alignment of reverse complements of a sequence. Obviously, there is only one permutation without inversions (namely the sorted sequence $(a_1 < a_2 < \cdots < a_n)$). Likewise, there is only one permutation with $\binom{n}{2}$ inversions (reverse-sorted sequence). Here $j$ and $k$ are integers (position of elements) whereas comparison takes place between the elements present at respective positions in a permutation. For instance, consider the set $S = \{D, E, C, A, B\}$ where each letter denotes an element. As $(D, E)$ is in proper order, it is not an inversion (as also $(A, B)$

in position $(4, 5)$). But all others $(D, C)$, $(D, A)$, $(D, B)$ are not in order so that the first 3 inversions are $(1, 3)$, $(1, 4)$, $(1, 5)$. Similarly, $(2, 3)$, $(2, 4)$, $(2, 5)$, $(3, 4)$, and $(3, 5)$ are also inversions. Thus, there are 8 inversions out of 10 possibilities. Clearly, the number of inversions depends on the permutation. It is a measure of the deviation or disagreement from a base case. This means that permutations with just one inversion is very close to the base case than those with more than one inversion. Permutations with more and more inversions deviate further away from the base case. This has applications in recommendation systems, computer sorting algorithms like bubble-sort and insertion-sort that tries to sort an arbitrary array by exchanging the data values that are "out-of-order" elements, and in computing Kendall's $\tau$.

A credible way to count the inversions is to use a divide-and-conquer strategy. First divide the permutation $P$ into $P1$ of size $n - m$ as $(a_1, a_2, \ldots, a_{n-m})$ and $P2$ $(a_{n-m+1}, a_{n-m+2}, \ldots, a_n)$ of size $m$. Then we count the inversions in $P1$ and $P2$, separately. These are called "intra-block-inversions." Afterward, we count the inversions of the form $(u, v)$ such that $u$ is in $P1$, and $v$ is in $P2$ and it is an inversion. This is called "inter-block-inversions." Counting the inversions is faster if the elements are re-ordered in "base-case order" (which means in increasing or decreasing order for numbers, and in alphabetic order for literals and strings). Inter-inversions can be skipped if both $P1$ and $P2$ are sorted, and the last element of $P1$ and first element of $P2$ does not form an inversion. In other words, there is no overlap between the boundaries of $P1$ and $P2$. As in the case of merge-sort, each sub-permutation ($P1$ and $P2$ initially) is further subdivided and the process repeated. Although this technique is useful to count the number of inversions, the following paragraph outlines a GF approach to the same problem.

Denote all permutations of $n$ elements with $k$ inversions by $I(n, k)$. For $k < n$, it satisfies the recurrence relation $I(n, k) = I(n - 1, k) + I(n, k - 1)$. The OGF for the sequence $I(n, k)$ is given as the polynomial product

$$F(t) = (1 + t)\left(1 + t + t^2\right)\left(1 + t + t^2 + t^3\right)\ldots\left(1 + t + t^2 + \cdots + t^{n-1}\right), \quad (4.1)$$

which in summed form is $\sum_{k=0}^{\binom{n}{2}} I(n, k)t^k$. Using the product notation, this can be expressed as $F(t) = \prod_{k=1}^{n-1}(1 + t + t^2 + \cdots + t^k)$.

### 4.1.3  Generating Function of Strided Sequences

Suppose the GF of a sequence $(a_0, a_1, a_2, \ldots, a_n, \ldots)$ is $F(t)$. It was shown in Chap. 2 that we could express the GF of odd and even terms (i.e., stride 2 terms) separately using $F(t)$. How do we find the OGF of a strided sequence of lag other than 2?. For example, what is the OGF of $\sum_{k \geq 0} a_{3k}t^{3k}$ in terms of $F(t)$? This can be found using $n^{th}$ roots of unity. Let

$\omega^n = 1$ define the $n^{th}$ roots of unity with $k^{th}$ root denoted by $\omega_k$ for $k = 0, 1, \ldots, n-1$. These are given by $\omega_k = \exp(2\pi i k/n)$ where "$i$" is the imaginary constant. If $k$ divides $n$, $\omega_j = 1$ for $j = 0, 1, 2, \ldots, (k-1)$. Otherwise $(1/k) \sum_{j=0}^{k-1} \omega_j = 0$. Thus, we have the relationship connecting the sum of the root of unity as $(1/k) \sum_{\omega^n=1}^{k-1} \omega_j = 1$ if $k$ divides $n$ and 0 otherwise. From this we get $F(t) = (1/k) \sum_{j=0}^{k-1} F(t\omega_j)$. This can be used to find the OGF of unambiguous context-free formal languages (Koutschan 2008) and Shapley-Shubik power indices of weighted voting games used in political science (Bilbao et al. 2000).

### 2. Generating function of strided sequences

Find the GF for $\sum_{k=0}^{\lceil n/3 \rceil} (-1)^k \binom{n}{3k}$ and obtain a compact expression for the sum, where $n$ is an integer and $k$ is less than $\lceil n/3 \rceil$.

Consider the cube roots of unity $1, \omega, \omega^2$. As the given sum involves binomial coefficients, we express the GF as $F(t) = (1+t)^n$. Using the above result, $G(t) = (1/3)[F(t) + F(t\omega) + F(t\omega^2)]$. To find a closed form expression for the given sum we put $t = -1$ in $G(t)$ to get $G(-1) = (1/3)[F(-1) + F(-\omega) + F(-\omega^2)]$. Denote these by $\omega_1 = (1 - \sqrt{3}i)/2$ and $\omega_2 = (1 + \sqrt{3}i)/2$. Substitute and simplify to get $3^{n/2}(2/3)\cos(n\pi/3)$ as the result.

## 4.2   Applications in Geometry

Consider the following simple problem. A circle is drawn arbitrarily on a plane initially. This divides the plane into one interior region (within the circle) and one exterior region (outside the circle). The process of drawing circles is repeated according to the following rule:– (i) each of the newly drawn circle should intersect already drawn circles at exactly two points (this means that tangential circles and interior circles are not allowed and any two circles cannot have the same center) (ii) three or more circles cannot pass through a common point of intersection. Each such newly drawn circle creates new regions to existing ones. Let $a_n$ represent the number of regions created by n circles. We could find a recurrence relation for $a_n$. Obviously $a_0 = 1$, because when no circles are drawn, we have only one region, namely the plane. When one circle is drawn, we have $a_1 = 2$. Similarly $a_2 = 4$, $a_3 = 8$, $a_4 = 14$, and so on (Fig. 4.1). The $n^{th}$ circle touches existing (n-1) circles in 2(n-1) points. Label these 2(n-1) points as $P_1, P_2, \cdots, P_{2(n-1)}$ in some fixed order (say clockwise or counterclockwise).[1] Note that these are all points on the $n^{th}$ circle. Each of the arcs $P_i$, $P_{i+1}$ on the $n^{th}$ circle creates a new region. This gives $a_n = a_{n-1} + 2(n-1)$, which is the recurrence sought for n≥2. Multiply both sides by $t^n$ and sum over 0 to $\infty$ to get

---

[1] Labels are chosen as roots of unity in some applications.

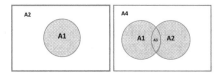

**Fig. 4.1** Intersecting circles

$$F(t) = tF(t) + 2\sum_{n=2}^{\infty}(n-1)t^n. \tag{4.2}$$

Write this in the form

$$(1-t)F(t) = 2t^2[1 + 2t + 3t^2 + \cdots] = 2t^2(1-t)^{-2}, \tag{4.3}$$

from which $F(t) = 2t^2/(1-t)^3$, which is the required OGF. The $n^{th}$ term is $2[1 + \binom{n}{n-2}] = 2 + 2\binom{n}{2} = 2 + n(n-1)$ where we have added a 2 because the second term on the RHS is valid for $n \geq 2$. The successive coefficients are 2, 4, 8, 14, 22, 32, 44, 58...

Another application of GF is in triangulation of a polygon with n sides. A polygon is a closed region with n line segments as sides. It is called a simple polygon if none of the edges intersect others, and a convex polygon if the line segment connecting any two interior points is always within it. Consider a convex polygon P with n+2 sides in $\mathbf{R}^2$. Triangulation is a technique to reduce a complex region into many simple non-overlapping triangles. Each vertex of the triangle is assumed to coincide with a vertex of the convex polygon in our discussion. It can be shown that the minimum number of triangles required for a simple n-gon is n-2. It has applications in computer vision, robotics, automatic mesh generation, etc. For example, it is used to place the minimum number of video cameras in a complex partitioned floor of a building, gallery, or factory in such a way that each nook and cranny is visible in at least one camera. The minimum number of cameras needed for a simple n-gon is $m = \lfloor n/3 \rfloor$. For instance, if the region is rectangular shaped, $m = \lfloor 4/3 \rfloor = 1$. If the camera has 90° visibility, it can be placed at any of the four corners, and if the visibility is 180°, it can be placed along a point on any side or on the roof.

Let $a_n$ denote the number of possible triangulations of polygon $P_{n+1}$ with noncrossing diagonals. Obviously $a_0 = 0, a_1 = a_2 = 1$. Label each of the nodes uniquely. There must be at least one triangle with its two vertices (say $u$ and $v$) coincident with adjacent vertices of the n-gon. Let the other vertex of the triangle be $w$. There are two cases to consider. If $w$ coincides with a neighboring vertex of the n-gon, we get a 2-partition of the n-gon (which can occur in two ways as $w$ may coincide with neighbor of either of the vertices of n-gon). Otherwise, the triangle will divide the n-gon into three parts. If we remove the common edge of the triangle and n-gon, we are left with either a smaller n-gon (with n-1 vertices), or two polygons with a common vertex. Depending upon the placement of $w$, there are $k$ vertices in one and $n + 1 - k$ vertices in the other for $k = 3, 4, \cdots, n - 2$. This gives the following recurrence

$$a_n = a_2 a_{n-1} + a_3 a_{n-2} + \cdots + a_{n-1} a_2. \tag{4.4}$$

This recurrence is already encountered in Chap. 1 on Catalan numbers. The solution is $C_{n-2} = \binom{2n-4}{n-2}/(n - 1)$. A related problem is enumerating the number of ways to place $m$ non-intersecting diagonals in a convex n-gon. The GF is also used to determine figurative numbers of geometric figures like tetrahedral, hexahedral, octahedral, dodecahedral, and icosahedral shapes (Carevic et al. 2020), and to enumerate the number of surfaces of regular n-simplices, and n-orthoplices in real dimensions (Lukaszyk 2022).

## 4.3 Applications in Computing

Computer algorithms are step-by-step methods to solve a practical problem using a computing device. Multiple methods may exist for solving the same problem. One common example is the sorting problem. Many algorithms (like quicksort, heapsort, mergesort, radixsort, insertion sort, bubble-sort, etc.) exist for data sorting. There are three primary concerns when choosing an appropriate algorithm (i) time complexity, (ii) space complexity, and (iii) communication complexity. Although some of the algorithms are easy to code, they may not have good performance for very large data sets. One example is bubble-sort algorithm which performs poorly when data size increases, but may be a reasonable choice when data size is always small (say $< 15$). This is where algorithm analysis steps in. If we could represent the complexity of an algorithm in terms of the input size, we could easily decide which algorithm to choose by comparing the complexity of different algorithms. The complexity of an algorithm is expressed as a mathematical function (in terms of the input size which is a number (size of data) in sorting algorithms, but is a function of a pair of numbers in 2D matrices (rows x columns) and graphs (nodes x edges)). Higher dimensional problems like 3D bin-packing, $n$-D matrix multiplications, etc. use more parameters.

### 4.3.1 Well-Formed Parentheses

Complex computer program statements in high-level languages are formed using parentheses. This not only reduces the program size but could result in speedy execution. Each left parenthesis appearing on the right side of executable statements (or on function calls, conditions used in 'if' and 'while' statements) should have a matching right parenthesis. As an example, $r1 = (-b + \text{sqrt}(b * b - 4 * a * c))/(2 * a)$ is a valid program statement with matching parentheses.[2] The number of well-formed parentheses can be enumerated using Catalan number $C(n)$ (which is discussed below). This can also be used in counting products (of real numbers, complex numbers, compatible matrices, etc.) around which parentheses can be put. Thus, if $A$, $B$, $C$ are compatible matrices, we could compute the product as $(A * B) * C$ or $A * (B * C)$. Similarly, laddered exponents (or repeated exponents) are expressions in which the exponent is repeated many times. Consider a two-level exponent $4^{3^2}$. It can be evaluated as either $4^{(3^2)} = 262144$ or as $(4^3)^2 = 4096$. This shows that braces can alter the meaning of a repeated exponent expression.

## 4.3.2  Merge-Sort Algorithm Analysis

The merge-sort algorithm uses the divide-and-conquer technique. The idea is very simple to understand. Suppose we need to arrange 64 (chosen for convenience, as it is a power of 2) school kids in increasing order of their heights. They are first divided arbitrarily into 2 groups of 32 each, say A and B. Each of the groups are further subdivided into 4 batches of 16 as A1, A2, B1, and B2. This process of sub-dividing goes on to get 8 batches of 8 each, 16 batches of 4 each, and finally 32 batches of 2 each (Fig. 4.2). Each of the subgroups must be uniquely identified. Now each group of 2 students are arranged in increasing order of height. Next they are regrouped into the same group from which they were separated, to get back 4 students in each batch. As they are already pair-wise arranged, it is a simple matter of merging the two sorted subgroups to get sorted quadruplets. This process of merging is continued until all 64 students are sorted by height. The last merging step involves two groups of 32 students in sorted order. This is the principle of merge-sort. If $C(n)$ denotes the complexity of arranging n entities ($n = 64$ in our problem) then

$$C(n) = C(\lceil n/2 \rceil) + C(\lfloor n/2 \rfloor) + k \lfloor n/2 \rfloor, \tag{4.5}$$

where $k$ is a constant close to 1 that takes care of sublist merging, and contingency costs (in case the merging is done by a separate function call), and n$\geq$ 2 (merge-sort requires

**Fig. 4.2** Mergesort tree

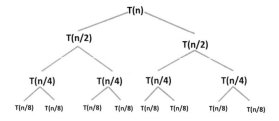

---

[2] Here sqrt is a library function (say in math or cmath library).

minimum 2 elements). This recurrence captures the average complexity of sorting n elements using mergesort, irrespective of whether the data are random or nearly sorted. To solve this recurrence, we assume that n is a power of 2.

Substitute $n = 2^m$ to get $C(2^m) = 2C(2^{m-1}) + k(2^{m-1})$ because both $\lceil 2^m/2 \rceil$ and $\lfloor 2^m/2 \rfloor$ are $2^{m-1}$. Divide both sides by $2^m$ to get $C(2^m)/2^m = C(2^{m-1})/2^{m-1} + k_1$ where $k_1 = k/2$. If $D(m)$ denotes $C(2^m)/2^m$, we could write this in terms of $D(m)$ as

$$D(m) = D(m-1) + k_1. \tag{4.6}$$

Multiply both sides by $t^m$, and add over $m = 1$ to $\infty$ to get $F(t) = \sum_{m=1}^{\infty} D(m)t^m = t \sum_{m=1}^{\infty} D(m-1)t^{m-1} + k_1 \sum_{m=1}^{\infty} t^m$. This becomes $F(t)[1-t] = k_1 t/(1-t)$. From this

$$F(t) = k_1 t/(1-t)^2. \tag{4.7}$$

The coefficient of $t^m$ is $k_1 m$. As we substituted $D(m) = C(2^m)/2^m$, we need to extract the coefficient of $n = 2^m$ (and not of $m$) in the resulting expression. This gives $C(n) = nk_1 \log_2(n)$. As $k_1$ is a constant close to 1, we get the complexity of merge-sort algorithm as $O(n \log_2(n))$ (see Knuth 1997; Sedgewick and Wayne 2011, and Sedgewick and Flajolet 2013). This gives us the following lemma.

---

**Lemma 4.1** *If the OGF of a nonlinearly transformed $(D(m) = C(2^m)/2^m)$ sequence $D(m) = D(m-1) + k$ is of the form $ct/(1-bt)^2$ where c is a small constant, and b is close to 1, then the complexity of the corresponding algorithm is $O(b^n n \log_2(n))$. Put $b = 1$ to get the average complexity of merge-sort algorithm as $O(n \log_2(n))$.*

---

### 4.3.3 Quick-Sort Algorithm Analysis

The quick-sort algorithm, invented by C.A.R. Hoare in 1962, is a popular sorting technique that uses divide-and-conquer strategy. It divides an array of size $n$ to be sorted into two subsets using a pivot (partitioning element) in such a way that all elements to the left of the pivot are less than (or equal) to it, and all elements to the right of it are greater than it (Fig. 4.3). This means that after successful completion of a *pass* through the data, the pivot will occupy its correct position so that it can be ignored in subsequent passes. This results in two subarrays whose total size is $(n-1)$. If they are not already sorted, we apply the same technique by choosing separate pivot elements from within their respective regions, and partitions them again. This process is continued until there are no more subarrays left for partitioning. The above discussion assumed that the pivot is an element of the array to be sorted. In fact, it can be any value (of course, of the same data type) within the minimum

**Fig. 4.3** Quicksort partitioning

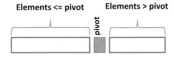

and maximum of the values to be sorted. As an example, it can be the average of 3 randomly chosen values for numeric data. For simplicity, it is assumed in the following discussion that the pivot need not be a value present in the array. If $C(n)$ denotes the average complexity of sorting n elements using quicksort,

$$C(n) = (n - 1) + (1/n) \sum_{k=1}^{n} (C(k - 1) + C(n - k)) \text{ for } n \geq 2. \tag{4.8}$$

The literal meaning of this expression is that we have to examine all the elements (except the pivot) with resulting complexity $(n - 1)$ to partition it using the pivot, and the resulting subarrays are of sizes $(k - 1)$ and $(n - k)$ where $k$ can take any of the values 1 through $n$. If $k = 1$, one of the subarrays will be of size 1, and the other of size $n - 1$ (which is a worst case). Similarly, if $k = n - 1$, the other sublist is of size 1. Ideally, we would like to partition in such a way that both subarrays are more or less of the same size. But this rarely happens in practice, unless the array to be sorted is already "nearly sorted," and the middle element is chosen as pivot.

Split Equation (4.8) into two, and put $j = n - k$ in the second term so that $j$ varies from $n - 1$ down to 0. Due to symmetry of the resulting sublists, we could write the above as $C(n) = (n - 1) + (2/n) \sum_{k=0}^{n-1} C(k)$. Multiply both sides by $n$ to get

$$n\,C(n) = n(n - 1) + 2 \sum_{k=0}^{n-1} C(k) \text{ for } n \geq 2. \tag{4.9}$$

Replace $n$ by $(n - 1)$ in (4.9), and subtract the resulting expression from it to get $nC(n) - (n - 1)C(n - 1) = 2n + 2C(n - 1)$ (where $-2$ is ignored). Rearrange to get $nC(n) = (n + 1)C(n - 1) + 2n$. Divide both sides by $n(n + 1)$. This gives

$$C(n)/(n + 1) = C(n - 1)/n + 2/(n + 1). \tag{4.10}$$

If $F(t)$ is the OGF of the sequence, we know from Chap. 2 that $C(n)/(n + 1)$ is the coefficient of $t^n$ in $\int F(t)dt$. Thus, we get $\int F(t)dt + C = t \int F(t)dt + C - 2\log(1 - t)$, because the first term on the RHS needs a right shift, and the last term is $2 \sum_{n=0}^{\infty} t^{n+1}/(n + 1) = 2 \int dt/(1 - t) = -2\log(1 - t)$ (Chap. 2). Ignoring constants we get $(1 - t) \int F(t)dt = -2\log(1 - t)$, from which $\int F(t)dt = -2\log(1 - t)/(1 - t)$. As the coefficient of $t^n$ in $\int F(t)dt$ is $a_{n-1}/n$, the time complexity of quicksort algorithm is expressible in terms of harmonic numbers as $O(nH_n)$, because the OGF of $H_n = 1 + 1/2 + \cdots + 1/n$ is $(1/(1-t))\log(1/(1-t))$, or equivalently $-\log(1-t)/(1-t)$. Although $O(nH_n)$ expresses the complexity succinctly, it is the usual practice to state it in terms of big-O notation as $O(n\log(n))$. Write $H_n$ as $\int_1^{n+1} dx/x$. This gives $H_n = \ln(n) + \gamma + \epsilon$ where $\gamma = 0.577215166$ is Euler's constant and $\epsilon$ is $O(1/(2n))$, which tends to zero for large $n$. Note that the logarithm is to the base $e$ in quicksort, and to the base 2 in mergesort (it matters asymptotically because $\log_e(n) = \log_2(n) \log_e(2) = 0.69314718 \log_2(n)$).

**Lemma 4.2** *If the integral of the OGF of a sequence is of the form $-c\log(1-t)/(1-t)$ where $c$ is a small constant, then the complexity of the corresponding algorithm is $O(n\log_e(n))$.*

### 4.3.4 Binary-Search Algorithm Analysis

The binary search is a popular algorithm for searching a key (data value) in a sorted array. It works by first finding the middle element (say mid) of the sorted array, and comparing the key with that element. The upper part (values greater than mid) could be totally ignored if the key is less than `mid`. On the other hand, if key is greater than mid, we could totally ignore the lower part. We call this part as the target block. The algorithm returns with a success code (key found) at any iteration when the mid element exactly matches the key. If not, the new middle element of the target block is found. The same process is repeated in each target block. This process is continued to bring down the size of the target part eventually to one. A success code is returned if the key matches that element. Otherwise, we report that the key is not found in the array. Let $C(n)$ denote the complexity of searching a key in a sorted array of size $n$. Then $C(n) = C(\lfloor n/2 \rfloor) + 1$, which is exact for odd $n$, and is an upper bound for even $n$. As done above, let $n = 2^m$, so that $C(\lfloor n/2 \rfloor) = C(\lfloor 2^{m-1} \rfloor)$. Let $B(m) = C(2^m)$ so that $B(m) = B(m-1) + 1$. Multiply both sides by $t^k$ and sum from 1 to $\infty$ to get the OGF as $F(t) = tF(t) + t/(1-t)$, from which $F(t) = t/(1-t)^2$. The coefficient of $t^m$ is $m = \log_2(n)$. This gives us the following lemma.

**Lemma 4.3** *If the OGF of a nonlinearly transformed recurrence $B(m) = C(2^m)$ is of the form $ct/(1-t)^2$ where $c$ is a small constant close to 1, then the complexity of the corresponding algorithm is $O(\log_2(n))$.*

An astute reader will notice that the OGF obtained for merge-sort and binary search are exactly identical (ie. $ct/(1-t)^2$). But the merge-sort used a nonlinear transformation $B(m) = C(2^m)/2^m$ whereas binary search used $B(m) = C(2^m)$. This is the reason why the complexity of merge-sort is $O(n\log(n))$ and that of binary search is $O(\lg(n))$.

### 4.3.5   Formal Languages

Consider a formal language $L$ accepted by a deterministic finite automata (DFA) say $M$ defined over a finite alphabet $\Sigma$. Label the start state as 0. Let $n_k$ denote the number of words in $L$ of length exactly $k$. This is the same as the number of directed paths of length exactly $k$ from 0 (start state) to the terminal state (any accepting state). Hence it must satisfy a linear recurrence relation, so that a GF exists for all such languages. This GF is a rational function (ratio of two polynomials with integer coefficients). In fact, there exists a relation between every formal power series with finite RoC and a formal regular language L if $a_n := |\{w \in L : |w| = n\}|$, where $G(x) = \sum_{n=0}^{\infty} a_n x^n$ where the $n^{th}$ coefficient gives the number of words of length n in L (Koutschan 2008). Let $N$ denote the transition matrix whose $(i, j)^{th}$ entry is the number of transitions (directed edges) from state $i$ to state $j$ of the automaton. If $a_n$ denotes the counting sequence of $L \subseteq \Sigma^*$, an OGF of $L$ can be defined as $F(t) = \sum_{n=0}^{\infty} a_n t^n$ where the coefficients $a_n = \#(x \in L||x| = n)$ (number of valid character strings whose length is $n$).[3] Finding the power series associated with a regular language is sometimes called Schützenberger method in automata theory. GFs are also used in many other fields of computer science. Time-complexity of any computational problem where the solutions to smaller non-overlapping sub-problem instances can be combined in an optimal way to get the solution to the original problem can be modeled using GFs. For instance, it is used to model the asymptotic resource complexity of functional programs, asymptotic enumeration problems, divide-and-conquer problems, etc.

## 4.4   Applications in Combinatorics

GFs are extensively used in enumeration problems. Some of them were discussed in Chaps. 1 and 2. In this section, we provide another application in proving combinatorial identities (Graham et al. 1994; Grimaldi 2019).

### 4.4.1   Combinatorial Identities

GFs are powerful tools to prove a variety of combinatorial identities. The tools and techniques developed in Chap. 2 can be used to easily prove some of the identities that may otherwise require meticulous arithmetic work. The inversion theorem $a_n = \sum_{k=0}^{n} \binom{n}{k} b_k \iff b_n = \sum_{k=0}^{n} (-1)^{n-k} \binom{n}{k} a_k$ where $a_0$ and $b_0$ are nonzero may also be needed in some problems.

---

[3] Identifier names in some languages are formed by string combinations in which some characters can't appear at certain positions.

**3. Prove $\sum_{k=0}^{n} 2^k \binom{n}{k} = 3^n$.**

Prove that $\sum_{k=0}^{n} 2^k \binom{n}{k} = 3^n$ for $n \geq 1$.

Consider the OGF of $2^k \binom{n}{k}$ as $F(x) = \sum_{k=0}^{n} 2^k x^k \binom{n}{k} = (2x + 1)^n$. Put $x = 1$ to get the result.

**4. Prove $\sum_{k=1}^{n} k\binom{n}{k} = n2^{n-1}$.**

Prove that $\sum_{k=1}^{n} k\binom{n}{k} = n2^{n-1}$ for $n \geq 1$.

We know that $\binom{n}{k}$ has GF $(1 + t)^n$ and multiplying an OGF by $1/(1 - t)$ results in the OGF of the partial sums of coefficients. In addition, the derivative of an OGF generates the sequence $ka_k$. Start with $F(t) = (1 + t)^n$. Then $F'(t) = n(1 + t)^{n-1}$ is the OGF of $k\binom{n}{k}$. Multiplying this by $1/(1 - t)$ results in the OGF of partial sums $s_j = \sum_{k=0}^{j} k\binom{n}{k}$ as $n(1 + t)^{n-1}/(1 - t)$. This is a convolution of respective OGFs so that $[x^n]\{n(1 + t)^{n-1}/(1 - t)\}$ is what we are looking for. By the convolution theorem in Chap. 2, this is $n \sum_{k=1}^{n-1} \binom{n-1}{k} = n2^{n-1}$.

**5. Prove $n \sum_{k=0}^{m-1} (-1)^k \binom{n}{k}$ is divisible by m.**

Prove that $n \sum_{k=0}^{m-1} (-1)^k \binom{n}{k}$ is divisible by both $m$ and $\binom{n}{m}$.

We know that $(-1)^k \binom{n}{k}$ has the OGF $(1 - x)^n$. We have already seen that the partial sums of these coefficients can be obtained by dividing this by $(1 - x)$. Thus, the OGF of the given sequence is the $(m - 1)^{th}$ term of $(1 - x)^n/(1 - x) = (1 - x)^{n-1}$. This is precisely $(-1)^m \binom{n-1}{m-1}$. Now consider $n\binom{n-1}{m-1} = n(n - 1)!/[(m - 1)!(n - m)!]$. Multiply numerator and denominator by m, write the denominator as $m!$, and write the numerator as $m * n!$ to get $m * n!/[m!(n - m)!] = m\binom{n}{m}$. This shows that the given sequence is divisible by both $m$ and $\binom{n}{m}$.

**6. Number of ways to choose elements from a set.**

If $S$ is a set of cardinality n, prove that the total number of ways to choose an even number of elements is equal to the total number of ways to choose an odd number of elements.

Mathematically this is equivalent to proving that $\binom{n}{0} + \binom{n}{2} + \binom{n}{4} + \cdots = \binom{n}{1} + \binom{n}{3} + \binom{n}{5} + \cdots = 2^{n-1}$ for $n \geq 1$. Add $\binom{n}{1} + \binom{n}{3} + \binom{n}{5} + \cdots$ to both sides so that the LHS becomes $\sum_{k=0}^{n} \binom{n}{k}$ which is easily seen to be $2^n$. The RHS is $2[\binom{n}{1} + \binom{n}{3} + \binom{n}{5} + \cdots]$. As the LHS is $2^n$ it follows that $\binom{n}{1} + \binom{n}{3} + \binom{n}{5} + \cdots = 2^{n-1}$.

**7. Prove $\binom{n}{0}^2 + \binom{n}{1}^2 + \cdots + \binom{n}{n}^2 = \binom{2n}{n}$.**

Prove that $\binom{n}{0}^2 + \binom{n}{1}^2 + \binom{n}{2}^2 + \cdots + \binom{n}{n}^2 = \binom{2n}{n}$ for $n \geq 1$.

As $\binom{n}{r} = \binom{n}{n-r}$, write the LHS as $\binom{n}{0}\binom{n}{n} + \binom{n}{1}\binom{n}{n-1} + \binom{n}{2}\binom{n}{n-2} + \cdots + \binom{n}{n}\binom{n}{0}$. This is the convolution of $(1+x)^n$ with $(x+1)^n$ which has OGF $(1+x)^{2n}$, which is generated by $a_n = \binom{2n}{n}$.

**8. Prove $\sum_{j=0}^{m} \binom{n-k}{j}\binom{k}{m-j} = \binom{n}{m}$.**

Prove that $\sum_{j=0}^{m} \binom{n-k}{j}\binom{k}{m-j} = \binom{n}{m}$ for every $k$ and $m \leq n$.

Obviously, the expression on the LHS is the coefficient of $m^{th}$ term in the convolution of two binomial terms $(1+t)^{n-k}$ and $(1+t)^k$. The OGF of the product as discussed in Chap. 2 has coefficient on the LHS. But as the product is $(1+t)^n$ which has $m^{th}$ term $\binom{n}{m}$. This proves the result.

**Problem 4.1** Use OGF to solve the recurrence $a_{n+1} = \frac{n+1}{n}a_n$, with $a_0 = 1$ (Hint: cross multiply to get $na_{n+1} = (n+1)a_n$).

**Problem 4.2** Use EGF to solve the recurrence $a_{n+1} = (n+1)a_n + n!$, with $a_0 = 0$ (Hint: Divide by $(n+1)!$ to get $a_{n+1}/(n+1)! = a_n/n! + 1/(n+1)$ and put $a_n/n! = b_n$).

**9. Prove $F(t) = \frac{1}{1-kt}G(\frac{t}{1-kt}) \Rightarrow G(t) = \frac{1}{1+kt}F(\frac{t}{1+kt})$.**

If $F(t)$ and $G(t)$ are two arbitrary GFs that satisfy the relationship $F(t) = \frac{1}{1-kt}G(\frac{t}{1-kt})$ then prove that $G(t) = \frac{1}{1+kt}F(\frac{t}{1+kt})$.

Change the variable as $s = t/(1-kt)$, so that $1 + ks = 1/(1-kt)$, and $t = s/(1+ks)$ to get $F(s/(1+ks)) = (1+ks)G(s)$. Divide both sides by $(1+ks)$, and replace $s$ by $t$ to get the result.

**10. Prove $\sum_{k=0}^{n} \binom{n}{k}/(k+1) = (2^{n+1}-1)/(n+1)$.**

Use the integral $\int_0^1 (1+x)^{n+1}dx = (2^{n+1}-1)/(n+1)$ to prove that $\sum_{k=0}^{n} \binom{n}{k}/(k+1) = (2^{n+1}-1)/(n+1)$.

Consider the binomial expansion $(1+x)^n = \sum_{k=0}^{n} \binom{n}{k}x^k$. Integrate both sides w.r.t. $x$ from 0–1 to get $\int_0^1 (1+x)^n dx = \sum_{k=0}^{n} \binom{n}{k} \int_0^1 x^k dx$. The LHS is $(1+x)^{n+1}/(n+1)|_0^1 = (2^{n+1}-1)/(n+1)$. The RHS is $\sum_{k=0}^{n} \binom{n}{k}/(k+1)x^{k+1}|_0^1$. When $x$ is 0, every term is zero, and when $x = 1$ we get the LHS of the above.

**11. Vandermonde identity.**

Prove Vandermonde identity $\sum_{j=0}^{k} \binom{m}{j}\binom{n}{k-j} = \binom{m+n}{k}$ using GFs.

We know that $\binom{m}{j}$ is the coefficient of $(1 + x)^m$ and $\binom{m+n}{k}$ is the coefficient of $(1 + x)^{m+n}$. Clearly, the LHS is the convolutions of two binomials. Write $(1 + x)^m (1 + x)^n = (1 + x)^{m+n}$. Expand the LHS and use the product of GF discussed in Chap. 2 to get the result.

### 4.4.2 New Generating Functions from Old

It was shown in Chap. 2 that a variety of transformations can be applied to existing GFs to get new ones. One simple case is to find the GF of even and odd terms. If $G(x) = a_0 + a_1x + a_2x^2 + \cdots + a_nx^n + \cdots$, we know that $(G(x) + G(-x))/2$ is the OGF of $a_0 + a_2x^2 + a_4x^4 + \cdots + a_{2n}x^{2n} + \cdots$ and $(G(x) - G(-x))/2$ is the OGF of $a_1x + a_3x^3 + a_5x^5 + \cdots + a_{2n+1}x^{2n+1} + \cdots$. This technique can be applied to any OGF or EGF.

**12.** $\sum_{k=0}^n a_k a_{n-k} = 1.$

Find the OGF and obtain an explicit expression for the $n^{th}$ term of a sequence with $a_0 = 1$ and $\sum_{k=0}^n a_k a_{n-k} = 1$.

The structure of the terms suggests that it must be a convolution (more precisely the square of an OGF). It was shown in Chap. 2 that the $n^{th}$ term of the square of an OGF is a convolution. Thus, we get $F(t)^2 = \sum_{n\geq 0} t^n = 1/(1 - t)$. Take square root to get $F(t) = (1 - t)^{-1/2}$ as the required OGF. This can be expanded as an infinite series $F(t) = \sum_{n\geq 0}(-1)^n\binom{-1/2}{n}t^n$. Hence, $a_n = (-1)^n\binom{-1/2}{n}$. This can be simplified as $a_n = (1.3.5.\ldots.(2n - 1))/(2^n n!)$. Particular values are $a_1 = 1/2$, $a_2 = 3/8$, $a_3 = 5/16$, etc.

**13. OGF of $b_n = \sum_{k=0}^n \binom{n}{k}a_k$.**

If $(a_0, a_1, \ldots,)$ is a given sequence, and $b_n = \sum_{k=0}^n \binom{n}{k}a_k$, find the OGF and EGF of the sequence $b_n$.

Let $F(t)$ denote the OGF of $(a_0, a_1, \ldots,)$, $G(t)$ denote the OGF of $b_n$, and H(t) denote the EGF of $b_n$. Then $G(t) = \sum_{n=0}^{\infty} b_n t^n$, and $H(t) = \sum_{n=0}^{\infty} b_n t^n/n!$. As the product of two OGF's has coefficient $c_k = \sum_{j=0}^k a_j b_{k-j}$, this is the convolution $1/(1 - t) * F(t/(1 - t))$. Substitute for $b_n$ to get $G(t) = \sum_{n=0}^{\infty} \sum_{k=0}^n \binom{n}{k}a_k \, t^n/n!$. As the product of two EGF's has coefficient $c_k = \sum_{j=0}^k \binom{k}{j}a_j b_{k-j}$, this is the convolution $e^t * F(t)$.

### 4.4.3 Recurrence Relations

Recurrence relations were briefly introduced in Chap. 1. They may be associated with a numerical sequence or functions with parameters. As examples, $F_n = F_{n-1} + F_{n-2}$ is a recurrence connecting three terms of a sequence, whereas $(x + 1)f(x; n + 1, p) = (n - $

$x) f(x; n, p)$ is a recurrence relation satisfied by the PMF of a binomial distribution with parameters $n$ and $p$. They are useful tools to find distribution functions, moments, and inverse moments of statistical distributions (Chattamvelli and Jones 1995). Unless otherwise stated, a recurrence relation will mean a relation connecting arbitrary terms of a sequence in the rest of this chapter.

**Definition 4.1** A recurrence relation is an algebraic expression that relates the $n^{th}$ term of an ordered sequence as a function of one or more prior terms along with a function of the index $n$.

Recurrence relations allows us to compute a new term of a sequence using one or more other known terms. By convention, the term with the highest index is expressed as a linear combination of prior terms. This allows forward computation of terms using known prior values, which may be stored in temporary arrays or simple variables (that are updated in each successive pass). The order of a recurrence relation is the difference between the maximum index, and minimum index in it. As an example, $a_n = 2a_{n-1} + 3^n$ is a first order nonlinear recurrence relation. Similarly, $F_n = F_{n-1} + F_{n-2}$ is linear of order 2 because $n^{th}$ term depends on $(n-2)^{th}$ term. The degree of a recurrence relation is the highest power involved in the coefficient. Thus, $a_n^2 = a_{n-1} * a_{n+1}$ is of degree two. Population growth models use the logistic map defined as $a_{n+1} = \lambda a_n(1 - a_n)$ where $0 < \lambda \leq 4$, and $|a_k| < 1$, which also is of degree two. A recurrence of degree 1 is a linear recurrence relation. If all coefficients are constants, it is called a recurrence relation with constant coefficients. Most of the examples discussed below are of this type. Homogeneous recurrences have no constant terms other than the coefficients. Thus, $F_n = F_{n-1} + F_{n-2}$ is a homogeneous recurrence, whereas $F_n = F_{n-1} + F_{n-2} + c$ is inhomogeneous. The characteristic equation (CE) of a recurrence is obtained by multiplying the $n^{th}$ term by $x^n$, summing over its range and identifying the multiplier of $F(x)$. Thus the CE of $F_n = F_{n-1} + F_{n-2}$ is $1 - x - x^2$. The OGF of a recurrence has the CE in the denominator (if common factors are absent in the numerator). The sign of a CE is irrelevant, but it does matter in the OGF. Recurrences can also be non-linear in the dummy variable(s). For example, $F(t) = t + F(t^2 + t^3)$ that counts the number of balanced 2–3 trees with $n$ external nodes is a non-linear recurrence in dummy variable $t$.

A set of initial conditions is necessary to solve it, depending on the order of the recurrence relation. A general recurrence of degree $k$ can be expressed as $c_1 a_n + c_2 a_{n-1} + \cdots + c_k a_{n-k} = f(n)$. There are many methods available for solving recurrence relations. This includes GFs, characteristic polynomial method, induction, substitution principle, table lookup methods, trial-and-error method, etc. Some of these may not be applicable in all problems. As an example, characteristic polynomial method is suitable when the terms involved in a recurrence are close-by. It may not be efficient in divide-and-conquer algorithms in which $C(n)$ is related to $C(n/2)$, as in the merge-sort algorithm (where the substitution method is ideal).

A closed form of a recurrence relation is an explicit function of the index variable. Consider the recurrence $a_n = 2a_{n-1} + 1$.

Multiply both sides by $t^n$ and sum over $n = 0$ to $\infty$ to get $\sum_{n=0}^{\infty} a_n t^n = 2 \sum_{n=1}^{\infty} a_{n-1} t^n + \sum_{n=0}^{\infty} t^n$. Take $t$ outside the summation from first term of the RHS to get $F(t) = 2t F(t) + 1/(1-t)$ or $F(t)[1 - 2t] = 1/(1-t)$, so that $F(t) = 1/[(1-2t)(1-t)]$. Write this as $A/(1-2t) + B/(1-t)$. Comparing coefficients give $A + B = 1$ and $A + 2B = 0$ from which $A = 2$ and $B = -1$. Thus, we get the OGF as $2/(1-2t) - 1/(1-t)$. Expand as an infinite series to get $a_n = 2^{n+1} - 1$.

Recurrence relations can be used to study the time complexity of algorithms. The idea is to express the complexity as a function of the input size $n$, and find a recurrence relation for this expression. This was already shown in Sect. 4.3.2 on mergesort, and Sect. 4.3.3 on quick-sort complexity.

**Problem 4.3** Find the OGF for the recurrence $a_{n+1} = 6a_n - 5a_{n-1}$ if $a_0 = 1, a_1 = 2$. Hint: Write $a_{n+1} - a_n = 5(a_n - a_{n-1})$, and put $b_{n+1} = a_{n+1} - a_n$.

**Problem 4.4** Find the OGF for Karatsuba n-digit integer multiplication algorithm with recurrence $T(n) = 3\ T(n/2) + n$, where $n \geq 2$, $T(0) = T(1) = 1$.

### 14. OGF of $\{a_n = n + 1\}$ for n odd.

Suppose a sequence $\{a_n\}$ is defined as $a_n = n + 1$ for n odd, and $a_n = 1$ otherwise. Find the OGF of $\{a_n\}$.

The given sequence is $\{1,2,1,4,1,6,\cdots\}$. The OGF is $F(t) = 1 + 2t + t^2 + 4t^3 + t^4 + 6t^5 + \cdots$. Split the odd and even terms to get the RHS as $(1 + t^2 + t^4 + \cdots) + (2t + 4t^3 + 6t^5 + \cdots)$. Put $t^2 = u$ to get the first part as $1/(1-u) = 1/(1-t^2)$. The second part is $2t(1 + 2t^2 + 3t^4 + \cdots) = 2t(1 - t^2)^{-2}$. Combine them to get $F(t) = 1/(1-t^2) + 2t/(1-t^2)^2$. Alternatively write $\{1,2,1,4,1,6,\cdots\} = \{1,2,3,4,5,6,\cdots\} - \{0,0,2,0,4,0,6,\cdots\}$, and use right-shifting property twice on second sequence. A slightly different, but equivalent expression is obtained by splitting it as $\{1,2,1,4,1,6,\cdots\} = \{1,1,1,\cdots\} + \{0,1,0,3,0,5,\cdots\}$. OGF of first one is $1/(1-t)$. Write the second one as $(t + 3t^3 + 5t^5 + \cdots) = \sum_{n=0}(2n+1)t^{2n+1}$. Write $t^{2n} = (t^2)^n$ to get the OGF as $t(1 + t^2)/(1 - t^2)^2$ (Example 3 in Chap. 1). Combining both gives the OGF as $1/(1-t) + t(1 + t^2)/(1 - t^2)^2$. Write $1/(1-t) = (1+t)/(1-t^2)$ and simplify to get above expression.

**15. OGF of $ca_n = ba_{n+1} + a_{n+2}$.**

Find OGF for the recurrence $ca_n = ba_{n+1} + a_{n+2}$, if $a_0 = 0$ and $a_1 = 1$.

Let $F(t)$ be the OGF. Then multiplying both sides by $t^n$, and summing over $n = 0$ to $\infty$ gives $cF(t) = b[F(t) - a_0]/t + [F(t) - a_0 - a_1 t]/t^2$. Solve for $F(t)$ to get $F(t) = a_0 + t(a_1 + a_0 b)/(1 + bt - ct^2)$. Put the initial values $a_0 = 0$ and $a_1 = 1$ to get $F(t) = t/(1 + bt - ct^2)$.

**16. OGF of $F_n/p^n$.**

Find the OGF for $F_n/p^n$ where $F_n$ is the $n^{th}$ Fibonacci number with $F_0 = 0$ and $F_1 = 1$, $p \neq 0$.

Let $F(t) = F_0 + F_1/p \, t + F_2/p^2 \, t^2 + \cdots + F_n/p^n \, t^n + \cdots$ be the OGF. Associate the $p^n$ in the denominator with the dummy variable as $(t/p)^n$ on the RHS. Proceed as in Example 15 in Chap. 1 to get $F(t)[1 - t/p - t^2/p^2] = F_0 + (F_1 - F_0)(t/p) +$ terms of the form $(F_k - F_{k-1} - F_{k-2})(t/p)^k$. All these coefficients vanish using Fibonacci recurrence, leaving behind $F(t)[1 - t/p - t^2/p^2] = F_0 + (F_1 - F_0)(t/p)$. Put $F_0 = 0$, and $F_1 = 1$ to get $F(t) = (t/p)/[1 - t/p - t^2/p^2] = pt/(p^2 - pt - t^2)$.

### 4.4.4   Towers of Hanoi Puzzle

The Towers of Hanoi (ToH) puzzle involves three identical vertical pegs (rods), and several discs of varying diameters that are all stacked upon an initial peg (source peg) in decreasing order of size. The challenge is to move all disks from the source peg (X) to the target peg (Z) using the other peg (Y) as an intermediary. The moves must satisfy the following conditions: (i) only one disk can be moved at a time (that is at the top of any peg) (ii) at no point in time should a disk of larger diameter be atop a smaller one on any of the pegs and (iii) only disks sitting at the top of any peg can be moved. There exists an elegant solution using recursion. Suppose we somehow move the top $n$-1 disks from the source peg X to the intermediate peg Y using target peg Z as an auxiliary. Then, it is a simple matter to move the largest diameter disk from source peg X to target peg Z. Now we have $n$-1 disks still sitting on the peg Y. If we re-label the source and intermediate pegs (exchange the role of X and Y), it should be possible to move $n$-1 disks from Y to Z using X as auxiliary. If $C(n)$ denotes the minimum number of moves needed to solve the puzzle, we get the recurrence $C(n) = 2C(n-1) + 1$, because $C(n-1)$ is the number of moves needed to first move the top $n$-1 disks from $X \mapsto Y$, and then from $Y \mapsto Z$. Obviously, $C(1) = 1$. Multiply both sides by $t^n$, and sum over n from 1 to $\infty$ (as the number of discs must be at least one) to get $\sum_{n=1}^{\infty} C(n)t^n = 2\sum_{n=1}^{\infty} C(n-1)t^n + \sum_{n=1}^{\infty} t^n$. Take one $t$ outside the first term of the RHS, and use $\sum_{n=1}^{\infty} t^n = t/(1-t)$ to get $F(t) = 2tF(t) + t/(1-t)$. This gives

$F(t) = t/[(1 - t)(1 - 2t)]$. Write this as A/(1–2t) + B/(1–t). Take (1-t)(1-2t) as a common denominator and compare the coefficients to get A + B = 0 and –2B–A = 1. This results in A = + 1 and B = − 1, so that

$$F(t) = 1/(1 - 2t) - 1/(1 - t). \tag{4.11}$$

The $n^{th}$ term is $C(n) = 2^n - 1$. For n = 1, 2, and 3 the number of moves are 1, 3 and 7 respectively.

## 4.5   Applications in Structural Engineering

Structural engineers design earthquake resistant structures under the assumption that oscillations occur in the direction of ground motion. Mathematical models can be developed to test the impact resistance of structures under cyclic loadings over time. Global structural vibrations can lead to local deformations and displacements due to impact vibrations at structurally weak sub-sections (or subframes). Impact forces and beam vibrations are modeled using differential or integral equations. Earthquake recurrence models approximate this type of nonlinear functions using linear recurrence relations (Bath 1978). Magnitude-recurrence relations can also be developed for various seismic events (quakes and aftershocks) and seismic statistical moments (Molnar 1979;  Atkinson and McCartney 2005). Perturbed linear systems are used to represent the response of the structure under horizontal translation as $g(x, t) = f(x(t), t) = \{f_k(x(t), t), k = 1, 2, \cdots, m\}$. Then $g(x, t)$ is analytical if $f_k(x(t), t) \forall k$ are so. This allows us to write $g(x, t) = \sum_{n=0}^{\infty} g^{(n)}(0)t^n/n!$, which is of EGF type. From this an EGF for $x(t) = \sum_{k=0}^{\infty} a_k t^k/k!$ could be obtained.

The GF approach can also be used for approximate analysis of the frames subjected to cyclic loads, and to get impact forces at particular time $t$ if the immediately previous force history is known.

## 4.6   Applications in Graph Theory

Graph theory is a branch of discrete mathematics in which inter-relationships (adjacency, incidence, containment, common traits, etc.) between entities are represented graphically, where nodes (vertices) represent entities, and edges (arcs) connecting them indicates relationships. A labeled graph is one with distinct labels assigned to the vertices. The labels can be numbers, letters, names, formulas (like chemical formulas), icons, or emojis, and the purpose is to uniquely identify them. Two labeled graphs are isomorphic if there is a one-to-one mapping between them (the nodes) that preserves the labels. A graph without self-loops (a mapping from a node to itself) is called a *simple* graph (otherwise it is multi-graph). All graphs in the following discussion are assumed to be simple graphs. A *connected* graph is one in which at least one path exists between any pair of nodes. Otherwise, it is called unconnected (or disconnected) graph. By convention, the null graph and singleton graph (with one node) are considered to be connected. Trees are special types of connected graphs in which there is just one path between any pair of nodes. A disconnected tree has a special name called *forest* (which is a collection of trees).

As a graph with $n$ vertices is a subset of the set of $\binom{n}{2}$ pairs of vertices, there exist a total number of $2^{\binom{n}{2}}$ graphs on n vertices. Hence, the GF for graphs with $n$ vertices is $G_n(x) = \sum_{k=0}^{\binom{n}{2}} G_{n,k} x^k$ where $G_{n,k}$ is the number of graphs with $n$ vertices and $k$ edges. Total number of simple labeled graphs with $n$ vertices and $m$ edges is $\binom{\binom{n}{2}}{m}$. This in turn results in the EGF $\sum_{n=0}^{\infty} 2^{\binom{n}{2}} x^n/n! = e^{C(x)}$ where $C(x)$ is the EGF of connected graphs.

### 4.6.1  Graph Enumeration

If there are n vertices available, any subset of the $\binom{n}{2}$ pairs of vertices (possible edges) should result in a graph. But all of them need not result in connected graphs. Consider all possible labeled graphs. As there are $\binom{n}{2}$ possible edges for $n$ nodes, there exist $\binom{\binom{n}{2}}{k}$ ways of choosing $k$ edges from those. Summing this over all possible values of $k$ gives the GF for the number of simple labeled graphs with $n$ nodes as $L_n(t) = \sum_{k=0}^{\binom{n}{2}} \binom{\binom{n}{2}}{k} t^k$. This has closed form $(1+t)^{\binom{n}{2}}$. Put $t = 1$ to get the total number of simple labeled graphs with $n$ nodes as $2^{\binom{n}{2}}$. This is the series A006125 in online encyclopedia of integer sequences (oeis.org/A006125). Let $C(n)$ denote the number of connected simple labeled graphs with $n$ nodes. It has the GF $G_n(t) = \sum_{k=0}^{\binom{n}{2}} G_{n,k} t^k$ where $G_{n,k}$ is the number of connected simple labeled graphs with $n$ vertices and $k$ edges. If this is divided into two sub-graphs with $k$ and $n - k$ components, it will satisfy the recurrence

$$C(n) = \sum_{k=1}^{n-1} \binom{n-2}{k-1} \left(2^k - 1\right) C_k C_{n-k} \quad \text{for} \quad n > 2, C(1) = 1. \tag{4.12}$$

This can be solved to get the recurrence relation

$$C(1) = C(2) = 1, C(n) = 2^{\binom{n}{2}} - n \sum_{k=0}^{n-2} \binom{n-1}{k} 2^{\binom{n-k-1}{2}} C_{k+1} \quad \text{for} \quad n > 2, \tag{4.13}$$

which is obtained by counting the total number of disconnected graphs and subtracting from $2^{\binom{n}{2}}$, being the total number of possible graphs (OEIS A001187). Note that $\binom{n-k-1}{2}$ is to be interpreted as zero when $n - k - 1$ is less than 2. Analogous results for directed graphs (digraphs) are obtained by replacing $\binom{n}{2} = n(n-1)/2$ by $n(n-1)$. Thus, there are $2^{n(n-1)}$ labeled digraphs on $n$ nodes. If $m > n(n-1)$, the total number of labeled digraphs with $m$ edges is the same as the number of digraphs with $n(n-1) - m$ edges. The GF is given by $G(t) = \sum_{k=0}^{n(n-1)} D_{n,k} t^k$ where $D_{n,k}$ is the number of directed graphs with $n$ nodes and $k$ edges (Stanley 1979).

Chromatic polynomials used in graph coloring is an alternating series which can be regarded as a Mobius function, or represented using Pochhammer falling factorial notation. For a complete graph $K_n$ with $n$ nodes, this takes the simple form

$$\chi_{K_n}(\lambda) = \lambda(\lambda - 1)(\lambda - 2)\ldots(\lambda - n + 1), \tag{4.14}$$

from which a falling Pochhammer GF for chromatic polynomial follows as

$$F_{K_n}(\lambda; t) = \sum_{n=1}^{\infty} (\lambda)_{(n)} t^n. \tag{4.15}$$

The distance between two vertices in a connected graph is the number of edges in the shortest path between them. Thus the distance between adjacent vertices is always 1. If $a_k$ is the number of pairs of vertices in a connected graph G that are at distance exactly k away, the OGF $W(t) = \sum_{k=0}^{d} a_k t^k$ where $d$ is the diameter (distance between farthest nodes) of G is known as Wiener polynomial in graph theory. The OGF of the distance sum follows as $W(t)/(1-t)$ (Chap. 1).

### Graph Enumeration Using EGF

The EGF is better suited because the number of graphs increases rapidly as a function of node size $n$. Let $C(t)$ denote the EGF for connected labeled graphs, and $D(t)$ denote the EGF for labeled graphs (which need not be connected). Then $D(t)$ is given by $D(t) = \sum_{n=0}^{\infty} 2^{\binom{n}{2}} t^n/n!$. They are related as $D(t) = \exp(C(t) - 1)$. This gives $C(t) = 1 + \ln(\sum_{n=0}^{\infty} 2^{n(n-1)/2} t^n/n!)$ where ln denotes log to the base $e$. Expand the infinite series to get $C(t) = 1 + \ln(1 + t + 2t^2/(2!) + 8t^3/(3!) + 64t^4/(4!) + \cdots)$. Now use $\log(1 + x) = x - x^2/2 + x^3/3 - \cdots$ to get $1 + t + t^2/(2!) + 4t^3/(3!) + 38t^4/(4!) + \cdots$. This shows that the number of connected graphs with k vertices is 1, 1, 1, 4, 38, etc., for $k = 0, 1, 2, \ldots$ (OEIS A001349).

## 4.6.2 Tree Enumeration

Enumerating trees can also be done easily using GFs. To illustrate, suppose we need to count the number of binary trees with $n$ vertices, or the number of ways in which a binary tree with $n$ nodes can be drawn on a plane (planar tree). This is given by the Catalan number discussed in Chap. 2 (Wilf 1994). Let $C(n)$ denote the number of binary trees with $n$ nodes. There is only one tree (a singleton node) when $n = 1$, so that $C(1) = 1$. Otherwise, count the number of nodes to the left of the root, and those to the right. If there are $j$ nodes in the left-subtree, there must be $n - 1 - j$ nodes in the right-subtree. As $j$ can vary between 0 and $n - 1$, we get $C(n) = C(0)C(n - 1) + C(1)C(n - 2) + \cdots + C(n - 1)C(0)$. This is the convolution which is already solved in Chap. 2. Thus, $C(n) = \binom{2n}{n}/(n + 1)$. The number of ordered binary trees on $n$ vertices is $C(n - 1)$. The total number of labeled trees on $n$ vertices is $n^{n-2}$. This is known as Caylay's formula. For rooted-trees this becomes $n^{n-1}$. The GF for the number of rooted trees with $n$ vertices is $r(t) = x \exp\left(\sum_{k=1}^{\infty} r\left(t^k\right)/k\right)$. The GF for counting general trees can be obtained from this using Otter's formula as $T(x) = r(x) - \left(r(x)^2 - r\left(x^2\right)\right)$.

There are many other interpretations to the Catalan number. Suppose an epidemic is spreading fast in a community. Males and females are equally likely to be infected. If it is known that exactly the same number of males and females have been infected, the cumulative number of male patients is at least as large as number of females is given by $C(n)$. Another application on matched-parentheses is discussed on Sect. 4.3.1. When two entirely different problems have the same GF, there must exist a one-to-one bijection between them. Thus, GFs are also used in other branches of science like theory of functions in mathematics, enumeration of organic compounds in chemistry, equivalent classes, etc.

## 4.7  Applications in Chemistry

Organic chemistry is the study of carbon compounds. An organic molecule is an assemblage of distinct atoms in which some atoms are linked to others by "valency bonds." Each single carbon atom has four valencies (four other atoms can be bonded to it). An *atom valence* in chemistry is the same as *vertex degree* in graph theory. The sum of the valencies of all vertices of a connected graph is twice the number of edges. This implies that the number of odd vertices in a connected graph is even. Molecular graphs (also called constitutional graphs) used in chemistry are graphs of chemical structures (the hydrogen atoms are usually suppressed as it can be inferred from the rest of the graph). Six bondings are possible when two carbon atoms form a direct link (called C–C link). Hydrocarbons are compounds that contain only carbon and hydrogen atoms (also called paraffins). Straight-chain (2D) hydrocarbons are called alkanes that have only C–C and C–H bonds. Chemical graph theory is a branch of mathematical chemistry that deals with applications of graph theory in chemistry (especially organic chemistry, where nodes represent atoms, and edges represent covalent two-electron bonds).

### 4.7.1   Polymer Chemistry

Polymers are high-molecular-weight organic compounds bonded or aggregated together by many smaller molecules called monomers. Several materials found in nature such as proteins, starch, synthetic fibers used in clothing, etc. are polymers. Plastics of various types, resins and rubbers are also polymers in wide use. Synthetic polymers account for more than half of the compounds produced by the chemical industry. A few monomer units are subjected to chemical reactions under controlled conditions to form polymers. The kinetic approach uses high-pressure tubular reactors that use a catalyst (usually oxygen) to bond the monomers. This is called addition reaction or chain growth. Another approach in which two or more molecules are combined, and a stable small molecule (like water or carbon dioxide) is eliminated, is called condensation reaction or step growth. Isomers are compounds that have identical formula but different structure or spatial arrangement. Working with polymers mathematically is a challenge due to the variety of structures. The PMF has been used for a long time to study them. The number, weight, and chromatographic probability mass functions are the most popular PGF in common use. The molecular weight distribution is usually assumed to follow the log-normal distribution. If the length of a molecule is assumed to follow a discrete distribution with PMF $P(N)$, the corresponding PGF is $F(t) = \sum_{n=0}^{\infty} t^n P(N = n)$.

**17. Number of polymers of length $n$**

Consider a polymer formed using 3 monomers say P, Q, R. Suppose P and Q can be used any number of times, but R can be used only an even number of times. Find the number of polymers of length $n$ that can be formed using P, Q, and R. We will use EGF for each of the possible choices. As P and Q can be used any number of times, the EGF for them is $\exp(t)$. As R can be used only an even number of times, it has EGF $\left(1 + t^2/2! + t^4/4! + t^6/6! + \cdots\right)$. As all possible choices are allowed, the EGF for our problem is the product $e^t e^t \left(1 + t^2/2! + t^4/4! + t^6/6! + \cdots\right)$. Using the result in Chap. 2, this can be simplified to $e^{2t} \left(e^t + e^{-t}\right)/2$. Split this into two terms to get $\left(e^{3t} + e^t\right)/2$. Expand each term as an infinite series, and collect the coefficient of $n^{th}$ term to get $a_n = (3^n + 1)/2$. If R can be used only an odd number of times, the corresponding EGF is $e^{2t} \left(e^t - e^{-t}\right)/2$. This gives $a_n = (3^n - 1)/2$. Both of them are integers because the last digit of $3^n$ is always 1, 3, 7, or 9, so that addition or subtraction of 1 from it results in an even integer, which when divided by 2 results in an integer.

Consider the problem of forming polymers using condensation approach, where one type of monomer is present. The OGF for the total number of ways in which this process can happen is $P_x(m, t) = \left[\sum_{k \geq 1} t^k\right]^m$, where $m$ is the total number of distinct polymers. Taking $t$ as a common factor, this can be written as $[t/(1 - t)]^m$ so that the coefficient $a_k = \binom{n-1}{k}$.

### 4.7.2    Counting Isomers of Hydrocarbons

Alkanes are hydrocarbons with general structure $C_nH_{2n+2}$, where C denotes Carbon atom, and H denotes Hydrogen atom, subscripts denote number of atoms of each type, and $n \geq 1$ is an integer.[4] Thus, for every carbon atom in an alkane, there are $2n + 2$ Hydrogen atoms (examples are $CH_4$ = Methane, $C_2H_6$ = Ethane, $C_3H_8$ = Propane, $C_4H_{10}$ = Butane), so that they are saturated (no free space exists for other atomic bonds). A tree ensues when every atom symbol (Carbon or Hydrogen) is replaced by a node, and every chemical bond by an undirected edge. Because of saturation, a tree representation (called molecular graph) assumes that carbon atoms have degree 4, and hydrogen atoms have degree 1. Alcohols are hydroxycarbons with general structure $C_nH_{2n+1}OH$. Thus a special type of tree called a "1–4 tree" is used to model such molecules. This tree has a property that each node has degree 1 or 4. A GF for alkanes (using unrooted 1–4 trees) is found in three steps. First find a GF for alcohols with $n$ carbon, $2n + 1$ hydrogen, and one OH group as

$$D(t) = 1 + t + t^2 + 2t^3 + 4t^4 + 8t^5 + 17t^6 + 39t^7 + 89t^8 + 211t^9 + \cdots, \qquad (4.16)$$

which satisfies the recurrence relation $D(t) = 1 + (t/6)\left[D(t)^3 + 3D\left(t^2\right)D(t) + 2D\left(t^3\right)\right]$. Next find a GF for rooted 1–4 trees with root as carbon atom (valency 4). This GF is given by

$$G(t) = t + t^2 + 2t^3 + 4t^4 + 9t^5 + 18t^6 + \cdots \qquad (4.17)$$

which satisfies the recurrence relation $G(t) = (t/24)\left[D(t)^4 + 6D(t)^2D\left(t^2\right) + 8\,D(t)^3D(t)\right.$ $\left. + 3D\left(t^2\right)^2 + 6D\left(t^4\right)\right]$. From this the GF for alkanes is found as

$$F(t) = 1 + t + t^2 + t^3 + 2t^4 + 3t^5 + 5t^6 + 9t^7 + 18t^8 + 35t^9 + 75t^{10} + 159t^{11} + \cdots \qquad (4.18)$$

which satisfy $F(t) = G(t) + D(t) - \left[D(t)^2 - D\left(t^2\right)\right]/2$.

There are no isomers for carbon count $n = 1, 2, 3$, but for $n = 4$, there exists two distinct isomers, and for $n = 5$ there are three isomers. The isomer count increases rapidly thereafter. Thus, there are 75 isomers for $n = 10$ and 366319 isomers for $n = 20$.

### 4.7.3    Modeling Polymerization

Polymerization is a process of reacting monomer molecules in a closed container as chemical reaction takes place to form 2D polymer chains or 3D chemical structure networks. There are two main classes of polymerization called "step-growth polymerization (SGP)" and "chain-

---

[4] Alkenes have double-bonds and have structure $C_nH_{2n}$ whereas Alkynes have triple-bonds and structure $C_nH_{2n-2}$.

growth polymerization (chain polymerization) (CGP)". Alkenes can easily form polymers through relatively simple radical reactions (eg: polyethylene and polyvinyl chloride (PVC)), whereas alkanes require strong acids as catalyst. SGP gets it's name from the fact that they are formed by independent and well-defined chemical reaction steps between functional groups of monomer units. CGP gets it's name because chain-extension reaction steps only occur while monomers are added to a growing chain with an active center. Radical polymerization at low conversions can be symbolically written as $s = sr$, and $r = e + mr$ where $s$ denotes a starter radical and $sr$ denotes a polymer radical, $e$ denotes an end-group, $m$ denotes a polymer with one or more monomers incorporated. If $m + e$ is assumed as 1, then $e$ can be regarded as probability of termination and $m$ as the probability of growth. Substitute for $r$ recursively to get $r = e/(1 - m)$. Now making the substitutions $m = m \exp(x)$, $e = e \exp(x)$ results in a MGF type expression (Chaps. 1, 2) $F(x) = se \exp(x)/[1 - m \exp(x)]$. This can be expanded as an infinite series $F(x) = \sum_n p(n) \exp(nx)$, which is the MGF because it is the expected value of $\exp(nx)$. The GF technique is a popular method in step-growth polymerization or modeling distributions of polymer properties (Asteasuain 2020). For example, the PGF method described in Chap. 3 can be used to find average of molecular weights, regulate the molecular weight distribution for polymerization processes (Zhang et al. 2022), find distribution of branched and cross-linked polymers, and of the degree of polymerization in the course of polymerization reactions and polycondensation reactions (Karl Fink 2014). The mathematical representation of CGP is

$$P^*(x) + M \rightarrow P^*(x + 1) + L \tag{4.19}$$

where $P^*()$ denotes a polymer at the active center of CGP, M is a monomer that reacts with active center, L is a by-product of chemical reaction obtained during chain propagation (which is usually a low-molar-mass chemical), $x = 1, 2, 3, \cdots$ represents the degree of polymerization. The OGF is also used in the theory of "monomer-dimer systems" of molecular chemistry and statistical physics.

## 4.8    Applications in Epidemiology

Spread of infectious diseases among multiple species in a community are modeled using mathematical techniques. Certain assumptions are often made in such modeling (transmission distribution is independent and identical, loop-backs do not exist, transmission is not rampant). For simplicity, it is assumed in the following discussion that a single species (humans) is involved (inter-species spread are not considered as in transmission through mosquitoes and insects). An outbreak indicates the beginning of an epidemic in a region at a certain time interval when one or more persons are infected from outside the population under study. A "contact" means that an infected person comes into contact with an uninfected person, who also gets the disease. This contact need not be physical (bodily), as there are diseases spread through contaminated water (Cholera), close physical proximity through air (COVID-19, TB, chickenpox, some viruses like SARS, Chikungunya), or sharing or handling of food (through an infected restaurant worker; e.g., Salmonella, Shigella, Campylobacter), items (like beds and sofas) or devices. There are also diseases that spread through multiple media (e.g., coronavirus through air, and intimate bodily contact). If a person is more likely to infect one of their contacts than to become infected, it is called asymmetric and otherwise symmetric. As an example, health-care workers (HCW), caregivers in community centers, and visitors are more likely to be infected by a patient, than otherwise. Whenever an individual transmits the disease to an uninfected one (offspring), it is captured into a graph, where nodes represent individuals, and an arc (edge) represents who infected whom. In other words, all individuals who comes into contact with an infected person (conditional contacts) are not part of the graph unless they also get infected. The edges can be directed or undirected. An undirected graph is used to model the occurrences rather than the direction of spread (for which directed graphs are used). Such a graph reduces to a tree when every offspring (patient) gets infected by a single source (person). Asymmetry may arise among males and females (as there are diseases with gender susceptibility, in which case two types of nodes are distinguished (say red and black or circle and square), or two separate graphs (one for each gender) are used). The spread can be one-to-one, one-to-many, or many-to-one (many-to-many are not considered as it can be modeled using many-to-one). Edge weights are probabilities (of transmission), contact duration, difference between ages of individuals, etc. Quite often, a special node (called start node that represents who originated the disease) is marked away to model the spread. Such epidemics are called single-source epidemic. But an epidemic may outbreak simultaneously with n individuals with absolutely no connectivity at all (multi-source epidemic).

One popular model used in epidemiology is the SIR model in which a population of $N$ individuals is divided into three mutually exclusive groups called Susceptible, Infected, and Removed (recovered (or dead)). Although dead persons cannot infect others, inclusion of them can throw more insight in some models. Moreover, in computer simulations, a dead person will necessitate a node and arcs removal (incident on the node) in the corresponding graph. This can easily be done if the graph is represented as an adjacency or incidence matrix. A person recovered from an infection is either returned back to the susceptible group (this is called SIS model which stands for Susceptible, Infected, Susceptible), or returned to a conditional group (some diseases like chickenpox and smallpox have a property that a recovered person will not get infected either throughout their lifetime, or during a fixed period). There are several quantities that can be modeled if the semi-directed degree distribution, and the probabilities of disease transmission to offspring are known. Examples are the number of active infections in a generation, number of recovered persons during a fixed time period, number of recovered and dead persons, proportion of infected population, etc.

Several mathematical models have been applied to capture the spread of COVID-19, and to come up with intervention measures. As the infection rate was very high in the beginning of the pandemic, a geometric model $T_N = (1 + m)^{N-1}$ where $T_N$ is the total new cases at time unit N and $m$ is the infectivity rate (average number of new infections by a single infected source person, which was later found to be time-dependent and not constant) was perceived to be appropriate. However, due to the rapid vaccination drives, people developed immunity against the coronavirus and the geometric growth rate diminished by a linear factor. COVID-19 modeling on directed graphs with heterogeneous transmission probabilities using GFs can be found in Widder and Schilling (2021). Several researchers have used differential equations to capture this rate of increase of COVID-19 patients (infected group) and rate of decrease of susceptible group. These were observed to be a better fit for real-world data at later stages of the pandemic.

The PGF approach is a mathematically sound method used in epidemiology and many other medical fields. It can be used to predict disease emergence, progression, and extinction. This has important implications in health care policy making, online monitoring and evaluation of disease progression, exceedences from known threshold transmission rates, developing effective plans of action, and in feeding information into early warning systems. Prior data on such epidemics may also be used to find the probabilities that an epidemic will last for a certain period of time before it is contained (brought under control), or in predicting patient deaths. To use the GF approach, it is assumed that the rate of transmission follows IID random variables. First consider a simple model that pertains to an infected individual. If $p_k$ denotes the probability that an individual (a node) infects $k$ others ($k$ offspring) during a fixed time interval (before recovery or death), we could model it using the PGF as $F(t) = \sum_{k \geq 0} p_k t^k$, where $p_0$ is the chance that an infected person does not transmit it to another. The expected number of infected individuals is then given by $F'(t)|_{t=1}$. If independence of random variables are assumed, all disease transmissions caused by an individual can be modeled using the geometric law. The PGF has closed form in such cases. As mentioned in Chap. 3, the probability that a single individual caused $k$ infections can then be obtained from the PGF as $p_k = (1/k!)(\partial/\partial t)^k F(t)|_{t=0}$. As shown in Chap. 3, they can also be used to find higher-order moments, and other statistics.

If the infected person transmits the disease to $m$ offspring (persons who contracts the disease from that person), we could consider each of them as independent events that follow the same distribution. Then the progress can be modeled as sums of IID random variables, whose coefficients are found using convolution (Chap. 2). As dividing an OGF by $(1 - t)$ results in an OGF with partial sums (CDFGF discussed in Chap. 3) as coefficients, we could estimate the probabilities of disease extinction and steadiness using PGF techniques. Assume that all individuals at some offspring generation stops further disease transmissions. In that case the probability of the convolutional sum will satisfy $p = P(t)|t = p$, and the complementary probability (disease established as a steady pandemic) is the solution of $1 - p = P(t)|t = p$. This is because $p\, P(t)/(1 - t)$ (the coefficients are $p$ times the CDF because $P(t)/(1 - t)$ is CDFGF) will result in the convolution of the coefficients of our OGF and $(1, 1, 1, \ldots)$, which when equated to $p$ will force the convolutional sum to be 1 so that "leftover probabilities" are zeros (which indicates that the disease spread stops). Our aim then is to estimate such a $p$. If $P(t)$ has closed form, this can be obtained either using analytical methods or iterative procedures.

   If the in-degrees (number of incoming edges to a node; $X$) and out-degrees (number of outgoing edges from a node; $Y$) are distinguished in a undirected network, the PGF becomes $P(s, t) = \sum_{j,k=0}^{\infty} p(j, k)s^j t^k$ where $p(j, k)$ is the joint PMF that a randomly chosen individual has $j$ contacts before being infected and produces $k$ offsprings (i.e., infects $k$ others who does not have the disease). Then the marginal PGF of $X$ (in-degree) is given by $P(s, 1)$, and that of $Y$ (out-degree) is given by $P(1, t)$. They are independent if $P(s, t) = P(s, 1)P(1, t)$ for all values of $s$ and $t$. If directed and undirected in- and out-degrees are separately considered, we need three dummy variables (one for incoming edges, two for outgoing directed and undirected edges). The incoming edges can also be considered separately as directed and undirected). An advantage of this approach is that the mean in-degrees (as in the case of HCW infections from patients), mean out-degrees (average number of persons to whom an infected person has transmitted the disease), etc., can be separately found out.

## 4.9   Applications in Number Theory

Number theory is a branch of mathematics that study various properties of numbers (usually positive integers). A partition of an integer $n$ is a sum of positive numbers that add up to $n$. A combinatorial argument can be used to find the number of partitions of $n$. Write $n$ as a linear list of $n$ number of 1's with a space in-between. For example $3 = 1\ 1\ 1$. Obviously, there are $n - 1$ spaces present. Let a slash character denote a division of the $n$ ones into groups. If there are $n - 1$ spaces, there exist $n - 1$ ways to place a single slash. This is the number of ways to select one slash from $n - 1$ which is $\binom{n-1}{1}$. Similarly, 2 slashes can be placed in $\binom{n-1}{2}$ ways, and so on. There is just one way to put $n - 1$ slashes in $n - 1$ spaces. Add them up to get $\binom{n-1}{1} + \binom{n-1}{2} + \cdots + \binom{n-1}{n-1}$ which is $(1 + 1)^{n-1} = 2^{n-1}$.

   Consider a positive integer $n > 1$. Let $p(n)$ denote the number of partitions where repetitions are allowed, and $Q(n)$ denote the partition into distinct parts. By convention, the elements are written in decreasing order (from largest to smallest). Obviously,

$p(0) = Q(0) = 1$. As parts can be repeated in $p(n)$, a 1 can appear any number of times from 0 up to $n$, so that it can be captured by the power series $1 + t + t^2 + t^3 + \cdots + t^n$. Similarly, the number of two's can be captured by $1 + t^2 + t^4 + \cdots$, and so on. Using the product rule, the OGF is the product $(1 + t + t^2 + t^3 + \cdots + t^n)(1 + t^2 + t^4 + \cdots) \ldots (1 + t^n)$. This is the same as $\sum_{n\geq0} p(n)t^n = \prod_{k=1}^{\infty} 1/(1 - t^k)$. This can also be written as $\sum_{n=1}^{\infty} p(n)t^n = 1 + \sum_{k=1}^{\infty} t^k/[(1 - t)(1 - t^2) \ldots (1 - t^k)]$. Similarly, the GF for $Q(n)$ is given by

$$\sum_{n\geq0} Q(n)t^n = 1 + \sum_{k=1}^{\infty} t^{\binom{k}{2}}/\left[(1 - t)(1 - t^2) \ldots \left(1 - t^k\right)\right]$$

$$= 1 + t + \sum_{k=2}^{\infty} t^k(1 + t) \ldots \left(1 + t^{k-1}\right)$$

which is convergent for $t < 1$. The GF can be used to prove that the total number of partitions of a positive integer $n$ into distinct parts is the same as the number of partitions of $n$ into odd parts. Here "odd parts" means that all parts are odd integers. For instance, $8 = 7 + 1 = 5 + 3 = 5 + 1 + 1 + 1$ contain only odd integers, whereas $8 = 6 + 2 = 4 + 2 + 2$ are "even partitions." Let $n$ be an integer $\geq 1$. The GF for the number of partitions of $n$ into distinct parts is $p(x) = \prod_{n\geq1}(1 + x^n)$. Now consider the partition of $n$ into odd parts. The OGF is $o(x) = \prod_{n\geq1}(1 - x^{2n-1})$. Use $(a + b)(a - b) = a^2 - b^2$, where $a = (1 - x^n)$ and $b = (1 + x^n)$ to get $(1 - x^{2n})/(1 - x^n) = (1 + x^n)$. Now $p(x) = \prod_{n\geq1}(1 + x^n) = \prod_{n\geq1}(1 - x^{2n})/(1 - x^n)$. Split this into the product of numerator terms, and product of denominator terms to get $p(x) = \prod_{n\geq1}(1 - x^{2n})/\prod_{n\geq1}(1 - x^n)$. Split the denominator product into "odd-terms product," and "even-terms product" as $\prod_{n\geq1}(1 - x^n) = \prod_{n\geq1}(1 - x^{2n-1})\prod_{n\geq1}(1 - x^{2n})$. Substitute in the above and cancel out $\prod_{n\geq1}(1 - x^{2n})$ from numerator and denominator to get $p(x) = 1/\prod_{n\geq1}(1 - x^{2n-1})$. This proves the result.

## 4.10 Applications in Statistics

Some of the applications of GFs in statistics appear in Chap. 3. Some more applications are introduced in this section.

### 4.10.1 Sums of IID Random Variables

Sums of independent random variables occur in many practical applications. The PGF in such cases is obtained using convolution, if they are assumed to be identically distributed. Consider a queuing model in which $X_k$ is the service time needed for customer $k$. If $n$

customers arrive in a fixed time period $t$, the total service time needed is $S_n = \sum_{k=1}^{n} X_k$. In an insurance domain, let $X_k$ denote the number of accidents, deaths, or claims received. If $n$ customers arrive in a fixed time period $t$, the total accidents, deaths, or claims received is $S_n = \sum_{k=1}^{n} X_k$. Similarly, let $n$ denote the number of customers who visit an ATM machine for cash withdrawal. If $X_k$ is the amount withdrawn by customer $k$ in a fixed time interval (say 24 h), then $S_n = \sum_{k=1}^{n} X_k$ is the total amount dispensed by the ATM.

If $X_1, X_2, \ldots, X_n$ are IID random variables with PGF $p_k(t)$, the sum $S_n = \sum_{k=1}^{n} X_k$ has PGF $P_{S_n}(t) = p_1(t) p_2(t) \ldots p_n(t)$. When all of them have the same distribution, we get $P_{S_n}(t) = [p(t)]^n$. If the PGF of the sum of a finite number of random variables is known, we could derive the distribution of the component random variables using independence and uniqueness assumptions.

**Theorem 4.1** *If $X_1, X_2, \ldots, X_N$ are IID random variables with PGF $G_{x_i}(t)$ where $N$ is an integer-valued random variable with PGF $F_N(t)$ which is independent of $X_i's$, then the sum $S_N = X_1 + X_2 + \cdots + X_N$ has PGF given by $H_{S_N}(t) = F_N(G_{x_i}(t))$.*

**Proof** By definition $H_{S_N}(t) = E\left(t^{S_N}\right) = E\left(E_{Nfixed}\left(t^{S_N}\right) | N\right)$. As $N$ is an integer valued random variable, this can be written as $E_{n \in N} E\left(t^{S_N} | N\right) P(N = n)$. Now expand $S_N$ to get $\sum_{n \in N} E\left(t^{X_1 + X_2 + \cdots + X_N}\right) P(N = n)$. This can be written as $\sum_{n \in N} \left[G_{x_i}(t)\right]^n P(N = n) = F_N\left(G_{x_i}(t)\right)$. This allows us to write $P S_N(x) = \sum_{n \in N} p_n P\left(S_n = y | N = n\right) = \left[t^x\right] \sum_{n \in N} p_n \left[G(t)\right]^n$. Differentiate w.r.t. $t$ and put $t = 0$ to get $E\left(S_N\right) = E(N)E(X)$.  □

### 18. OGF of heads in a fair die toss

A fair die with faces marked $1 - 6$ is thrown, and the number $N$ that turns up is noted. Then a coin with probability of head is $p$ is tossed $N$ times. Find the OGF for the number of heads that turns up.

We know that the probability of getting $N$ is $1/6$ as the die is fair, where $N = 1, 2, \ldots, 6$. This gives the OGF as $P(t) = \sum_{k=1}^{6} t^k / 6$. As the second part involves Bernoulli trials $N$ times, we have a BINO$(N, p)$ distribution with PGF $(q + pt)^N$. As $N$ is a random variable, we find the PGF of the number of heads as $\sum_{n=1}^{6} p_n [q(t)]^N = (1/6) \left[\sum_{n=1}^{6} (q + pt)^n\right]$.

## 4.10.2 Infinite Divisibility

A statistical distribution $X$ is called *infinite divisible* iff for each positive integer $n$, it can be represented as a sum of n IID random variables $S_n = X_1 + X_2 + \cdots + X_n$, where each of the $X_k$'s have a common distribution. Additivity and infinite divisibility are related because some distributions are infinite divisible using location parameters. One common example is the normal law $N\left(\mu_1, \sigma_1^2\right) + N\left(\mu_2, \sigma_2^2\right) = N\left(\mu_1 + \mu_2, \sigma_1^2 + \sigma_2^2\right)$, if they are independent. Other stable distributions (e.g., Cauchy distribution) are also infinitely divisible by location parameter. The characteristic function of such distributions have an inherent property that it is the $n^{th}$ power of some characteristic function due to the fact that the characteristic function of a sum of IID random variables is the product of individual characteristic functions. Symbolically, $\phi(t) = [\phi_n(t)]^n$. But the converse need not be true. Consider the binomial distribution with characteristic function $(q + pe^{it})^n$. If $n$ is finite, we could write this as a convolution only in a finite number of ways because $n$ must also be positive. As an illustration of infinite divisibility, the characteristic function of a Poisson distribution with parameter $\lambda$ can be written as $\exp\left[\lambda\left(e^{it} - 1\right)\right] = \left(\exp\left[\lambda/n\left(e^{it} - 1\right)\right]\right)^{1/n}$, showing that the Poisson distribution is infinitely divisible. Steutel and Van Harn (1979) proved that a distribution with non-negative support (i.e., it takes values $0, 1, 2, \ldots$) with $p_0 > 0$ is infinitely divisible iff it satisfies

$$(n + 1)p_{n+1} = \sum_{k=0}^{n} q_k p_{n-k}, \quad \text{for } n = 0, 1, 2, \ldots, \tag{4.20}$$

where $q_k$ are non-negative, and $\sum_{k=0}^{\infty} q_k/(k + 1) < \infty$. As the RHS represents the convolution $\left(p_0 + p_1 t + p_2 t^2 + \cdots + p_n t^n + \cdots\right)\left(q_0 + q_1 t + q_2 t^2 + \cdots + q_n t^n + \cdots\right)$, and $(n + 1)p_{n+1}$ is the $(n + 1)^{th}$ term of $(t\partial/\partial t)F(t)$ where $F(t)$ is the PGF of the distribution, we could state the above as follows: "A distribution with PGF $F(t)$ is infinitely divisible if $(t\partial/\partial t)F(t)$ can be represented as a convolution of $F(t)$ and $G(t)$, where $G(t)$ is the PGF of $q_k$'s."

### 4.10.3 Applications in Stochastic Processes

Consider a population of organisms that undergoes a birth-death process. The model is purely deterministic if the reproduction occurs at a constant rate. In practice, however, the reproduction is random as many extraneous factors may affect it. Such processes are called stochastic processes. Assume that h is a short period of time during which reproduction occurs with probability $\lambda h$. Time is measured in discrete units. This will of course depend on the organism under study. The size of the population in current generation $t$ determines its size in the next generation $(t + h)$. There exist two possibilities if the population count is to reach $N$ at time $(t + h)$. Either it is $N$ at time $t$, and no birth occurs during the time interval $h$, or it is $(N - 1)$ at time $t$ and 1 birth occurs during the interval $h$. Probability of more than 1 birth occurring during interval $h$ is precluded by choosing $h$ small enough for each species. Hence, the probability of $(N - 1)$ species to increase in size to $N$ in time interval $(t, t + h)$ is $\lambda(N - 1)h$. Likewise, the probability of $N$ species to increase in size to $N + 1$ in time interval $(t, t + h)$ is $\lambda Nh$. Then the complementary probability $1 - \lambda Nh$ will denote probability of no increase. Denote by $p_N(t)$ the probability that the population is of size $N$ at time $t$. Then

$$p_N(t + h) = p_N(t)(1 - \lambda Nh) + p_{N-1}(t)\lambda(N - 1)h. \tag{4.21}$$

Divide both sides by $h$ and take the limit $h \to 0$ to get

$$Lt_{h \to 0} \left[ p_N(t + h) - p_N(t) \right] / h = \partial p_N(t) / \partial t = -\lambda N p_N(t) + \lambda(N - 1) p_{N-1}(t). \tag{4.22}$$

The solution is $p_N(t) = \binom{N-1}{n_0} \exp(-\lambda n_0 t) (1 - \exp(-\lambda t))^{N-n_0}$ where $n_0 = N(0)$. This is the negative binomial distribution discussed in Chap. 3. As it is a special case of the geometric distribution, we get $p_N(t) = \exp(-\lambda t)(1 - \exp(-\lambda t))^{N-1}$ when $n_0 = 1$.

Consider a discrete branching process where the size of the $n^{th}$ generation is given by $X_n = \sum_{k=0}^{X_{n-1}} Y_k$, where $Y_k$ is the offspring distribution of $k^{th}$ individual. If all $Y_k$'s are assumed to be IID, it is easy to see that

$$P\left[ X_n = k | X_{n-1} = j \right] = P\left[ Y_1 + Y_2 + \cdots + Y_j \right]. \tag{4.23}$$

Using total probability law of conditional probability, this becomes

$$P\left[ X_n = k \right] = \sum_{j=0}^{\infty} P\left[ Y_1 + Y_2 + \cdots + Y_j \right] P\left[ X_{n-1} = j \right]. \tag{4.24}$$

Let $\phi_n(t)$ denote the PGF of the size of $n^{th}$ generation. Then

$$\phi_n(t) = \sum_{k=0}^{X_{n-1}} P(X_n = k) t^k. \tag{4.25}$$

From this it follows that $\phi_n(t) = \phi_{n-1}(\phi(t))$. Jormakka and Ghosh (2021) applied PGF to a birth-death stochastic process of M/M/1 queuing system and obtained steady-state probabilities for time-independent Markov process yielding polynomial time algorithms for certain partitions. They also derived a theoretical polynomial-time algorithm to the knapsack problem.

## 4.11  Applications in Reliability

A component in a system is considered as either working, or non-working in reliability theory. Let the probability that it is working be $p$. Then $q = 1 - p$ denotes the probability of a non-working component. This is called a binary system. An extension in which each component is perfectly working, partially working, and non-working is called a multi-state system. A system with $n$ identical components[5] that works only if $k$ out of $n$ of them works is called a $k$-out-of-$n$ system. Obviously, $k$ is less than $n$. One example is in internal combustion engines with identical spark-plugs. An automobile with six spark plugs may work properly when two of them are down. Some machines may work with a lower performance when the number of components further goes down (as in automobiles or multi-processor based parallel computers), but there is a limit on the number of non-working components in electronics, computer networking, etc. It is assumed that all components are independently and identically distributed in practical modeling of $k$-out-of-$n$ systems, because it greatly simplifies the underlying mathematics. The probability that the number of working components is at least $k$ is given by

$$R(k, n) = \sum_{j=k}^{n} \binom{n}{j} p^j q^{n-j}. \tag{4.26}$$

This represents the survival probability of a binomial distribution BINO$(n, p)$. A GF for this could be obtained using the SFGF discussed in Chap. 2. Replace $k$ by $k + 1$ in (4.26) and subtract from it to get the recurrence $R(k, n) - R(k + 1, n) = \binom{n}{k} p^k q^{n-k}$. This has a representation in terms of the incomplete beta function. Using the symmetry property of beta distribution, (4.26) can also be represented as

$$R(k, n) = p^k \sum_{j=k}^{n} \binom{j - 1}{k - 1} q^{j-k}. \tag{4.27}$$

Several variants of this system are in use in various disciplines. As an example, some machine learning ensemble-models are built using several (say $n$) base models. If $m$ among the base models have error rate $\epsilon < 1/2$, we could get an estimate of the error made by the ensemble-model using independence of base models. Suppose each model is used for data classification. Then $P[\text{ensemble error}] = P[m/2 \text{ or more models misclassify data}] = \sum_{k=m/2}^{n} \binom{n}{k} p^k (1 - p)^{n-k}$, where $p$ denotes the probability of error made by each of the base models. As another example, consider the "$n$-version" systems that are mission critical software, firmware or hardware systems independently developed by different groups of persons who do not communicate among each other. Different groups may use different tools and technologies (like programming languages or operating systems in $n$-version software system or processors from different manufacturers in $n$-version hardware systems). The systems are tested simultaneously, and a voting mechanism is used to decide which of the

---

[5] Theoretically the components need not be exactly identical as in distributed computing or network routing.

$n$ systems produced the best output. Assume that all versions have the same reliability $r$. If majority-voting mechanism is used, the reliability can be expressed as

$$R(k, n) = \sum_{j=\lceil k \rceil}^{n} \binom{n}{j} r^j (1 - r)^{n-j} \tag{4.28}$$

which is structurally similar to Eq. (4.27). The same technique is also used in fault-tolerant software programs where different software versions are studied for failure events that are assumed to be independent.

### 4.11.1 Series-Parallel Systems

Suppose a system is composed of several parts in series. This means that the parts are connected in series in the case of hardware systems. But in the case of software systems, this means that there are sequential modules to be executed one after another. Assume that each part is comprised of $m_i$ components in parallel. Such parallel components increase the reliability of the system because if one of them becomes non-functional, the others will spring into action, and keep the system going. In the case of software systems, this may mean that the same functionality is implemented using several "functions" that use different algorithms or programming languages. If one function throws an exception (recoverable run-time error), we could call one of the other identical functions. Such systems are called series-parallel systems. Let $r_i$ denotes the reliability of component $D_i$. If there are $n$ components in series, the reliability of the entire system is $\prod_{k=1}^{n} r_k$. As each series component is comprised of $m_i$ parallel components, the reliability at stage $i$ is $(1 - r_i)^{m_i}$. Assume for simplicity that each of the $r_i$ are the same $(r)$. Then the reliability of the entire system is $\prod_{k=1}^{n} (1 - r)^{m_k}$. To find the GF for this expression, we take log to get $\log(F(t)) = \sum_{k=1}^{n} (1 - r)^{m_k} t^k$. This has closed form expression when either each of the $m_i$'s are the same, or they are distributed according to one of the well-known series.

### 4.12   Applications in Bioinformatics

GFs are used in various places in bioinformatics. As an example, suppose there are restrictions on how many of each type of $(A, C, G, T)$ (where $A =$ Adenine, $C =$ Cytosine, $G =$ Guanine, and $T =$ Thymine) are allowed in a protein combination. These can be incorporated into a GF as described in what follows.

### 4.12.1 Lifetime of Cellular Proteins

Assume that the lifetime of cellular proteins is exponentially distributed. If there are $n$ independent protein molecules under observation, the mean time for protein degradation is the mean of the minimum of $n$ IID exponential random variables, which is $1/(n\lambda)$, because the CDF of the minimum is $1 - \exp(-n\lambda x)$. Let $T_n$ denote the time until the final protein degrades. Then $T_n = \sum_{k=1}^{n} 1/(k\lambda)$. Take $1/\lambda$ as a constant outside. This shows that the OGF is $F(t) = (1/\lambda) \sum_{k=1}^{n} t^k/k = (1/\lambda)H_n$ for $t = 1$, where $H_n$ is the harmonic number, already discussed above. GFs are also used for enumerating the number of $k$-noncrossing RNA structures.

### 4.12.2 Sequence Alignment

Sequence alignment is a popular technique in proteomics, bio-informatics, and genomics. A string will denote a sequence of letters from a fixed alphabet. Position-specific scoring matrix (PSSM), also called a profile weight matrix is often used in these fields (e.g., amino acids in proteomics, or nucleotide sequences within DNA in genomics) to capture the information contained in multiple alignments. Let $W$ denote a weight matrix where $n$ columns contain alignment data, and $m$ rows contain the letters of the alphabet. Then $w_{ij}$ denotes the score for letter of type $a_i$ and in column $j$. As OGF often uses positive coefficients, the PSSM score may have to be scaled to positive real numbers. GFs used in bioinformatics and genomics are of the weighted-type in which the dummy variable (Chap. 1) is raised to $w_{ij}$ to get the weighted OGF (WOGF)

$$F_j(t) = \sum_{k=1}^{m} p_k t^{w_{kj}}, \tag{4.29}$$

where $p_k$ is the probability assigned to the distribution of letters in the string, and the alignment of sequences are assumed to be IID random variables. Assuming independence of letters, the WOGF for PSSM match is given by the product of individual OGFs as

$$F(t) = \prod_{j=1}^{n} \sum_{k=1}^{m} p_k t^{w_{kj}}. \tag{4.30}$$

The coefficient of $t^s$ gives the probability of getting a score $s$ when $F(t)$ is expanded as a power series.

## 4.13 Applications in Genetics

Genomes are made up of chromosomes in which the smallest distinguishable unit is a gene. Several living species have very similar gene arrangements, but the order tends to differ. Clusters of living organisms can be arranged as a multiway tree based on the extend of overlap in their genomes. Computational genetics deals with the study of similarities and differences among the genes of living organisms. There are several genetic operations like mutation (changing a single gene position in a chromosome), cross-over (that involve two chromosomes), inversion, etc. (Chattamvelli 2016). The PGF can be used for enumerating the number of crossovers. Mapping functions used in genetics are mathematical functions to quantify the relative positions of genetic markers arranged linearly on a map. This avoids the problem of recombination fractions not being additive for two-locus models. The earliest known mapping function is called the Haldane map (1919) $m = f(x) = -\log(1 - 2x)/2$ for $0 \le x < 0.5$, which has the inverse map $x = (1 - \exp(-2|m|))/2$. If $M(x)$ denotes a mapping of physical distances to genetic distances, $F(t)$ denotes the PGF of the distribution of crossovers, $\mu$ denotes the mean number of crossovers, and $p_k$ denotes the probability of $k$ crossovers, we could define $M(x)$ in terms of PGF as

$$M(x) = (1 - F(1 - 2x/\mu))/2. \tag{4.31}$$

Assume that the number of crossovers follow the binomial distribution $BINO(n, p)$. Then from Chap. 3, $F(t) = (q + pt)^n$ where $q = 1 - p$. Substitute $t = 1 - 2x/\mu$ where $\mu = np$, and cancel out $p$, and use $p + q = 1$ to get the PGF as $F(1 - 2x/\mu) = (1 - 2x/n)^n$. Substitute in (4.31) to get the map $M(x) = \left[1 - (1 - 2x/n)^n\right]/2$ with $M(0) = 0$. Putting $n = 1$ gives $M(x) = x$. This is called complete interference, as there is only one possible crossover. As $\lim_{n \to \infty}(1 - a/n)^n = \exp(-a)$, the above expression approaches the well-known Haldane map. Simplified expression can be obtained when the number of crossovers are odd or even, as discussed in Chattamvelli and Shanmugam (2020). Probability of more than $k$ or less than $k$ crossovers, where $k$ is a fixed constant $< n/2$ can also be found. A zero-truncated binomial distribution is used when the probability of "no crossover" is to be ignored. Catalan numbers discussed in Chap. 2 also finds applications in genomics. If crossovers are restricted to adjacent positions only, the number of possible ways in which single crossovers could occur is given by the Catalan number.

A recurrence for RNA configurations was reported by Howell et al. (1980) as

$$R(n + 1) = R(n) + \sum_{j=1}^{n-1} R(j)R(n - j), \quad \text{for } n \ge 2, \tag{4.32}$$

with R(0) = R(1)= R(2) = 1. This was extended by Waterman (1995), Hofacker et al. (1998) to RNA secondary structures, who obtained the recurrence

$$S(n + 1) = S(n) + \sum_{j=M}^{n-1} S(j)S(n - j - 1), \quad \text{for } n \geq m + 1, \tag{4.33}$$

with $S(0) = S(1) = S(2) = \cdots S(m - 1) = 1$ (See Pages 36–38, Chap. 2).

Many diversity measures and indices are also used in genomics. One simple example is the gene diversity measure $H$ defined as $C * H = 1 - \sum_{k=1}^{m} x_k^2$, where $n$ is the number of individuals in a population, $m$ is the number of alleles at a locus, $x_k$ is the frequency of $k^{th}$ allele, and $C$ is either $(1 - 1/n)$ or $(1 - 1/(2n))$ depending on whether the species under consideration is self-fertilizing or not. The multiplier $C$ acts as a small-sample correction factor. When all alleles have equal frequency, we have $x_k = 1/m$ so that $1 - \sum_{k=1}^{m} x_k^2 = 1 - \sum_{k=1}^{m}(1/m)^2 = 1 - m/m^2 = (1 - 1/m)$. Recurrences can easily be developed for $n$ fixed and $m$ varying, or *vice versa*. This reduces to the heterozygosity average[6] when $C = 1$.

Recurrences are also used in equilibrium proportion for adjacent generations, and in inbreeding models of genetics. For example, when alleles are drawn randomly from $N$ individuals in a population, the probability that the same allele is drawn twice is $1/(2N)$, and the two alleles drawn are different is $1 - 1/(2N)$. If $f_t$ denotes the inbreeding coefficient for generation $t$, the probability that two alleles are identical by descent satisfy the recurrence $f_{t+1} = C + (1 - C)f_t$, where $C = 1/(2N)$. Subtract both LHS and RHS from 1 to get $1 - f_{t+1} = (1 - C)(1 - f_t)$. Put $(1 - f_t) = g_t$ to get $g_{t+1} = (1 - C)g_t$, which is a geometric progression (GP) with common factor $1 - C$, so that it can be solved as in Chap. 2 to get $g_t = (1 - C)^t g_0$, where $g_0$ is the initial heterozygosity. Substitute back $g_t = (1 - f_t)$ to get $(1 - f_t) = (1 - C)^t(1 - f_0)$. An EGF and a recurrence relation for computing the number of topologically distinct clone orderings can be found in Newberg (1996), and GFs to count the number of signed and unsigned arrangements of ordered type in genomics can be found in Tesler (2008). Multivariate extensions of OGF with applications in run-statistics can be found in Kong (2019).

## 4.14  Applications in Management

GFs are not very popular in management science, as in other fields (exceptions are in pricing and dynamic market models discussed below). This is perhaps due to the fact that the majority of GFs encountered in management sciences are simple in structure, for which first- or second-order recurrence relationships exist. This is precisely the reason for the popularity

---

[6] Different genes that occupy the same locus are called alleles, and individuals in a population with two different alleles for some gene is called heterozygous for that gene.

of recurrences in banking and finance, accounting, portfolio management, pricing dynamics, and other fields. A first-order recurrence relation relates an arbitrary term (other than the initial one) in a sequence to the adjacent (previous or next) term in the same sequence. It is called linear if it does not involve square-roots, fractional powers or higher-order terms. In other words, it is of the form $A_{n+1} = CA_n + D$ where $C$ and $D$ are constants. The initial term $A_0$ must be known to start the iterations. A first-order linear recurrence is used in flat-rate, unit-cost, and reducing-balance depreciation models. As an example, the simple interest can be represented as a first-order recurrence $A_{n+1} = A_n + Pr/100$ if $r$ is the annual interest rate in percentages, and as $A_{n+1} = A_n + Pr/12$ if $r$ is a fraction $(0 - 1)$, and interest is computed monthly. The divisor must be carefully chosen depending upon whether the interest rate is specified in percentage or as a fraction, and the period is yearly, half-yearly, quarterly or monthly. Similarly, flat rate and unit cost depreciations are linear recurrences $A_{n+1} = A_n - Pr/100$ where $P$ is the initial value of an asset, $A_{n+1}$ is asset value after $n$ years. These are called linear growth and decay models.

A geometric sequence is a set of ordered numbers (real or complex) in which each term (except the first) is obtained by multiplying the previous term by a fixed number, called the common ratio. It can be finite or infinite. Quite often, they are finite in management sciences. Expressions involving GPs appear at different places in economics and finance. One example is in geometric growth and deprecation models. Deprecation is the reduction in value over time of an asset. Mathematically $d_n = v_{n-1} - v_n$ where $v_n < v_{n-1}$ because of wear and tear of the asset or value fluctuations over time. This is a first order linear recurrence. This can be approximated using a GP for some of the assets. For instance, if the deprecation rate is constant (say 0.90), we could write this as $d_n = (0.90)^{n-1} v_0 - (0.90)^n v_0 = (0.90)^{n-1}(0.10) v_0$ which is in GP. Similarly, return on investments (RoI) of growing assets or funds are given by $RoI(n) = v_n - v_{n-1}$, which is the difference in the value of the asset from year $n - 1$ to year $n$.

If $P$ is the principal amount deposited in a compound interest scheme that yields $r$ percent interest rate per annum, the accumulated amount at the end of $n$ years is given by $A = P(1 + r/100)^n$. The corresponding recurrences for geometric growth and decay models are $A_{n+1} = A_n R$ where $R = (1 + r/100)$ for compound interest, and $r$ is the interest rate per compounding period. The balance depreciation models[7] are similar, except that $R = (1 - r/100)$. If $A_k$ denotes the accumulated amount at the end of $k^{th}$ year, an OGF for accumulated amount can be found as $F(t) = \sum_{k=1}^{n} A_k t^k$. For convenience, it is assumed that $n$ is very large (tends to $\infty$) because the GF method does not care even if the sequence goes to infinity as our aim is just to extract one or two coefficients from it. Substitute for $A_k$, and take $P$ as a constant to get $F(t) = P \sum_{k=0}^{\infty} (1 + r/100)^k t^k = P/[1 - (1 + r/100)t]$.

---

[7] This also applies to value depreciation models of assets or items that are acquired for use, and undergoes wear-and-tear or decrease in quality due to continuous use or inactivity.

This is a compact expression which is amenable to further operations. When this is expanded as an infinite series, the coefficient of $t^n$ will give the accumulated amount at the end of $n$ years. Suppose we wish to determine how much is the growth from $m^{th}$ year to $n^{th}$ year where $m < n$. This is given by $[t^n]\{P/[1 - (1 + r/100)t]\} - [t^m]\{P/[1 - (1 + r/100)t]\}$. It satisfies the recurrence relationship $A_n = (1 + r/100)A_{n-1}$ with $A_0 = P$.

The linear and geometric models may have to be combined in some problems. Consider a situation where the interest on a loan is compounded continuously but loan repayments are made on a regular basis. This results in a linearly decreasing and geometrically increasing model, which will reduce the balance if the repayment amount is greater than the interest. This can be modeled as $A_{n+1} = (1 + r/100)A_n - C$, where $C$ is the annual amount repaid. Replace $(1 + r/100)$ by $(1 - r/100)$, and $-C$ by $+C$ to get a geometrically decaying, and linearly increasing model. Both the linear and geometric terms can also be increasing in some problems. Consider an investment scheme in which a fixed amount $C$ is deposited every period (say monthly), and interest on current balance is also compounded during the same period. This results in the recurrence $A_{n+1} = (1 + r/100)A_n + C$ where $A_0$ is the first payment (principal amount, which can be $C$). Similarly, if the *monthly* payment on a loan of initial amount $P$ is $C$, the balance $A_{n+1}$ satisfies the recurrence $A_{n+1} = (1 + r/(12x100))A_n - C$ where $A_0 = P$.

### 4.14.1 Annuity

An annuity is a mortgage or trust account into which regular payments are made. The recurrences for annuity and pensions where a certain amount is invested, and regularly withdrawn (usually after a lapse of time) follow the same pattern. Equal amounts of cash or other asset flows occurring as a stream at fixed periods is known as annuity, and if the flows are unequal, it is called a mixed stream (where $C$ will vary). Consider an annuity with fixed interest rate $r$ into which a constant amount $C$ is invested. The future value ($FV$) at the *end of $n$* periods is given by

$$FV_n = C\left[R^n + R^{n-1} + \cdots + R^1\right], \tag{4.34}$$

where $R$ is $(1 + r)$ or $(1 + r/100)$ as discussed above, and it is assumed that payments are made in the *beginning* of each period (so that interest is accrued on it by the end of the period). Note that the initial amount $C$ becomes $CR$ at the end of the period, so that initial $C$ is not counted. Take $R$ as common in RHS of (4.34), and use the sum of a GP to get $FV_n = CR\left[(R^n - 1)/(R - 1)\right]$. This satisfies the recurrence $FV_n = FV_{n-1} + CR^n$. The net present value (NPV) models (also known as discounted cash flow models) are very similar because $R$ is used as divisor.

$$NPV = \left[ C_n/R^n + C_{n-1}/R^{n-1} + \cdots + C_1/R + C_0 \right], \tag{4.35}$$

where $C_0, C_1, \ldots, C_n$ are the cash inflows during time periods 0 through $n$, and interest is compounded.

## 4.15  Applications in Economics

Economy growth and decay models are used to model the future growth of world economies. This is either applied at entity levels (like a country), sector (like agriculture or industrial output of various industries), or commodity level. The simplest one is the geometric model discussed above as $E_{n+1} = RE_n$, with initial condition $E_0 = K$, where $E_0$ denotes a particular stage in the economy. This may be either the beginning of a year, before an economic recession sets in, or may coincide with another related event (like natural disasters, economic reforms, new rules or laws that affect economy, changes in governance due to elections, etc.). Here $R$ is an average $\sum_{i=1}^{n} r_i/n$, computed from key economy factors like inflation rate, import-export deficits (balance of trade), foreign exchange rates of various currencies, etc. The magnitude of $R$ decides whether growth or decay occurs. A GF can easily be obtained as before. But $R$ need not remain a constant over time. Thus, recurrence relations are much better suited to model, as they allow $R$ to vary over time. A combined linear and geometric model can also be used when inflows (like exports, loans, bank contributions) and outflows (imports, loan repayments) are incorporated. The GF approach can be used when derivatives (say of first- or second-order) are involved on either side of the equation, as in pricing models. As discussed in Chap. 2, derivatives of GFs results in a left-shift of the coefficients.

## 4.16  Miscellaneous Applications

GFs are used in many other applied science fields. For example, auto-covariance identification among signals and white noise in DSP uses GFs to filter signal components. Forecasting models also use it for stationary time series data filtering. GF of transition probabilities in homogeneous continuous time birth-and-death process is used for analysis or approximation of steady state probabilities. Optimal feedback control of spacecrafts expending continuous thrust using Hamilton-Jacoby theory with split boundary conditions can be modeled using GFs as well  (Park et al. 2006).  Okundamiya and Ojieabu (2010) used GFs for the performance evaluation of communication systems for effective data transmission. See  Lee et al. (2013) for accurate numerical scheme for solving PGFs arising in stochastic models of general first-order reaction networks by using the characteristic curves.

## 4.17 Summary

Several GFs have important roles in the analytical study of polynomials in real and complex variables. This chapter discussed several applications of GFs in algebra, analysis of algorithms, combinatorics, economics, epidemiology, genetics, graph theory, management, number theory, organic chemistry, reliability theory, and statistics. They are powerful tools in the analysis of algorithms in which complexity is expressed in terms of recurrence relations. It is shown (in Page 83) that the quick-sort algorithm runs around thirty percent faster than merge-sort algorithm for large input size (n). This chapter also discussed several recurrence relationships encountered by researchers and professionals in many scientific disciplines.

## References

Asteasuain, M. (2020). Efficient modeling of distributions of polymer properties using probability generating functions and parallel computing. *Computers and Chemical Engineering, 128*(2), 261–284. (sciencedirect.com). https://doi.org/10.1016/j.compchemeng.2019.06.009

Atkinson, G. M., & McCartney, S. E. (2005). A revised magnitude-recurrence relation for shallow crustal earthquakes in southwestern British Columbia: Considering the relationships between moment magnitude and regional magnitudes. *Bulletin of the Seismological Society of America, 95*(1), 334–340. https://doi.org/10.1785/0120040095

Bath, M. (1978). A note on recurrence relations for earthquakes. *Tectonophysics, 51*(1, 2), T23–T30. https://doi.org/10.1016/0040-19517890047-1

Bilbao, J. M., Fernández, J. R., Posada, A. J., & López, J. J. (2000). Generating functions for computing power indices efficiently. *Top, 8,* 191–213. https://doi.org/10.1007/BF02628555

Bona, M. (2012). *Combinatorics of permutations* (2nd edn.) CRC Press.

Carevic, M. M., Petrovic, M. J., & Denic, N. (2020). Generating function for the figurative numbers of regular polyhedron. *Mathematical Problems in Engineering, 2020,* Article ID 6238934, 1–7. https://doi.org/10.1155/2020/6238934

Castor, C. A., Sarmoria, C. A, Asteasuain, C. M, Brandolin, A., & Pinto, J. C. (2014). Mathematical modeling of molecular weight distributions in Vinyl Chloride suspension polymerizations performed with a bifunctional initiator through Probability Generating Functions (PGF). *Macromolecular Theory and Simulations, 23*(8), 500–522. https://doi.org/10.1002/mats.201400038

Chattamvelli, R. (2016). *Data mining methods*. Oxford, UK: Alpha Science.

Chattamvelli, R., & Jones, M. C. (1995). Recurrence relations for noncentral density, distribution functions, and inverse moments. *Journal of Statistical Computation and Simulation, 52*(3), 289–299. https://doi.org/10.1080/00949659508811679

Chattamvelli, R., & Shanmugam, R. (2020). *Discrete distributions in engineering and the applied sciences*. Springer.

Dobrushkin, V. A., & Sahni, S. (2009). *Methods in algorithmic analysis*. CRC Press.

Fertin, G., et al. (2009). *Combinatorics of genome rearrangements*. Cambridge, Massachusetts: MIT Press.

Fontana, W., Konings, D. A. M., Stadler, P. F., & Schuster, P. (1993). Statistics of RNA secondary structures. *Biopolymers, 33,* 1389–1404.

Fortunatti, C., Sarmoria, C., Brandolin, A., & Asteasuain, M. (2014). Modeling of RAFT polymerization using probability generating functions. *Macromolecular Reaction Engineering, 8*(12), 781–795. https://doi.org/10.1002/mren.201400020

Graham, R., Knuth, D. E., & Patashnik, O. (1994). *Concrete mathematics* (2nd ed.). MA: Addison Wesley.

Grimaldi, R. P. (2019). *Discrete and combinatorial mathematics: An applied introduction* (5th ed.). Pearson Education.

Gutman, I., & Polansky, O. E. (1986). *Mathematical concepts in organic chemistry*. Berlin: Springer.

Hartleb, D., Ahrens, A., Purvinis, O., & Zascerinska, J. (2020). Analysis of free time intervals between buyers at cash register using generating functions. In *Proceedings of 10th international conference on pervasive and parallel computing, communications and sensors- PECCS* (pp. 42–49).

Hofacker, I. L., Schuster, P., & Stadlerab, P. F. (1998). Combinatorics of RNA secondary structures. *Discrete Applied Mathematics, 88*(1–3), 207–237. https://doi.org/10.1016/S0166-218X9800073-0

Howell, J. A., Smith, T. F., & Waterman, M. S. (1980). Computation of generating functions for biological molecules. *SIAM Journal of Applied Mathematics, 39*(1), 119–133.

Jormakka, J., & Ghosh, S. (2021). Applications of generating functions to stochastic processes and to the complexity of the Knapsack problem, Preprints 2021, 2021040706. https://doi.org/10.20944/preprints202104.0706.v1

Karl Fink, J. (2014). *Generating functions in polymer science, chapter 17 of physical chemistry in depth* (pp. 443–485). Springer. ISBN-13: 978-3642424403.

Knuth, D. E. (1997). *The art of computer programming* (Vol. 1). Reading, MA: Addison-Wesley.

Kong, Y. (2019). Generating function methods for run and scan statistics. In Glaz, J., & Koutras, M. V. (Eds.), *Handbook of Scan Statistics*. Springer. https://doi.org/10.1007/978-1-4614-8414-1_56-1

Koutschan, C. (2008). Regular languages and their generating functions: The inverse problem. *Theoretical Computer Science, 391*, 65–74. sciencedirect.com. https://doi.org/10.1016/j.tcs.2007.10.031

Lee, C. H., Shin, J., & Kim, J. (2013). A numerical characteristic method for probability generating functions on stochastic first-order reaction networks. *Journal of Mathematical Chemistry, 51*, 316–337. https://doi.org/10.1007/s10910-012-0085-8

Lukaszyk, S. (2022). Novel recurrence relations for volumes and surfaces of n-Balls, Regular n-Simplices, and n-Orthoplices in real dimensions. *MDPI in Mathematics, 10*. https://www.mdpi.com/2227-7390/10/13/2212/pdf. https://doi.org/10.3390/math10132212

Molnar, P. (1979). Earthquake recurrence intervals and plate tectonics. *Bulletin of the Seismological Society of America, 69*(1), 115–133.

Newberg, L. A. (1996). The number of clone orderings. *Discrete Applied Mathematics, 69*(3), 233–245. https://doi.org/10.1016/0166-218X9600093-5

Okundamiya, M. S., & Ojieabu, C. E. (2010). Performance analysis and evaluation of communication systems. *Journal of Mobile Communication, 4*, 47–53. https://doi.org/10.3923/jmcomm.2010.47.53

Park, C., Guibout, V., & Scheeres, D. J. (2006). Solving optimal continuous thrust rendezvous problems with generating functions. *Journal of Guidance, Control, and Dynamics, 29*(2), 321–342. https://doi.org/10.2514/1.14580

Riordan, J. (1962). Generating function for powers of Fibonacci numbers. *Duke Mathematical Journal, 29*, 5–12.

Riordan, J. (1979). *Combinatorial identities*. New York: Wiley.

Scranton, A. B., Klier, J., & Peppas, N. A. (1991). Statistical analysis of free-radical copolymerization/crosslinking reactions using probability generating functions: Reaction directionality and general termination. *Macromolecules, 24*(6), 1412–1415. American Chemical Society. https://doi.org/10.1021/ma00006a031

Sedgewick, R., & Wayne, K. (2011). *Algorithms*. MA: Addison-Wesley.

Sedgewick, R., & Flajolet, P. (2013). *An introduction to the analysis of algorithms*. MA: Addison-Wesley.

Stanley, R. P. (1979). *Algebraic combinatorics: Walks trees: Tableaux, and more*. Springer.

Steutel, F. W., & Van Harn, K. (1979). Discrete analogues of self-decomposability and stability. *Annals of Probability, 7*, 893–899. https://doi.org/10.1214/aop/1176994950

Tesler, G. (2008). Distribution of segment lengths in genome rearrangements. *Electronic Journal of Combinatorics, 15*, #R105.

Waterman, M. S. (1995). *Introduction to computational biology: Maps, sequences, and genomes*. London: Chapman & Hall.

Widder, C., & Schilling, T. (2021). Generating functions for message passing on weighted networks: Directed bond percolation and susceptible, infected, recovered epidemics. *Physical Review - E, 104*, 054305. https://journals.aps.org/pre/pdf/10.1103/PhysRevE.104.054305

Wilf, H. (1994). *Generating functionology*. Academic Press. https://www2.math.upenn.edu/~wilf/gfology2.pdf

Zhang, J., Pu, J., & Ren, M. (2022). Molecular weight distribution control for polymerization processes based on the moment-generating function. *Entropy, 24*(499), 499 (2022). https://doaj.org/article/12e98397b3704d5cb016dbffaf13e6c0. https://doi.org/10.3390/e24040499

Zhang, J., Shen, F., & Waguespack, Y. (2016). Incorporating generating functions to computational science education. In *International conference on computational science and computational intelligence (CSCI)* (pp. 315–320). Las Vegas.

# Index

R. Chattamvelli and R. Shanmugam, *Generating Functions in Engineering and the Applied
Sciences*, Synthesis Lectures on Engineering, Science, and Technology,
https://doi.org/10.1007/978-3-031-21143-0

Printed in the United States
by Baker & Taylor Publisher Services